Towards Analytical Chaotic Evolutions in Brusselators

Synthesis Lectures on Mechanical Engineering

Synthesis Lectures on Mechanical Engineering series publishes 60–150 page publications pertaining to this diverse discipline of mechanical engineering. The series presents Lectures written for an audience of researchers, industry engineers, undergraduate and graduate students.

Additional Synthesis series will be developed covering key areas within mechanical engineering.

Towards Analytical Chaotic Evolutions in Brusselators
Albert C.J. Luo and Siyu Guo
2020

Modeling and Simulation of Nanofluid Flow Problems
Snehashi Chakraverty and Uddhaba Biswal
2020

Modeling and Simulation of Mechatronic Systems using Simscape
Shuvra Das
2020

Automatic Flight Control Systems
Mohammad Sadraey
2020

Bifurcation Dynamics of a Damped Parametric Pendulum
Yu Guo and Albert C.J. Luo
2019

Reliability-Based Mechanical Design, Volume 2: Component under Cyclic Load and Dimension Design with Required Reliability
Xiaobin Le
2019

Reliability-Based Mechanical Design, Volume 1: Component under Static Load
Xiaobin Le
2019

Solving Practical Engineering Mechanics Problems: Statics
Sayavur I. Bakhtiyarov
2017

Unmanned Aircraft Design: A Review of Fundamentals
Mohammad Sadraey
2017

Introduction to Refrigeration and Air Conditioning Systems: Theory and Applications
Allan Kirkpatrick
2017

Resistance Spot Welding: Fundamentals and Applications for the Automotive Industry
Menachem Kimchi and David H. Phillips
2017

MEMS Barometers Toward Vertical Position Detection: Background Theory, System
Prototyping, and Measurement Analysis
Dimosthenis E. Bolanakis
2017

Engineering Finite Element Analysis
Ramana M. Pidaparti
2017

Towards Analytical Chaotic Evolutions in Brusselators

Albert C.J. Luo and Siyu Guo

ISBN: 978-3-031-79660-9   paperback
ISBN: 978-3-031-79661-6   ebook
ISBN: 978-3-031-79662-3   hardcover

DOI 10.1007/978-3-031-79661-6

A Publication in the Springer series
SYNTHESIS LECTURES ON MECHANICAL ENGINEERING

Lecture #27
Series ISSN
Print 2573-3168   Electronic 2573-3176

Towards Analytical Chaotic Evolutions in Brusselators

Albert C.J. Luo and Siyu Guo
Southern Illinois University Edwardsville

SYNTHESIS LECTURES ON MECHANICAL ENGINEERING #27

ABSTRACT

The Brusselator is a mathematical model for autocatalytic reaction, which was proposed by Ilya Prigogine and his collaborators at the Université Libre de Bruxelles. The dynamics of the Brusselator gives an oscillating reaction mechanism for an autocatalytic, oscillating chemical reaction. The Brusselator is a slow-fast oscillating chemical reaction system. The traditional analytical methods cannot provide analytical solutions of such slow-fast oscillating reaction, and numerical simulations cannot provide a full picture of periodic evolutions in the Brusselator. In this book, the generalized harmonic balance methods are employed for analytical solutions of periodic evolutions of the Brusselator with a harmonic diffusion. The bifurcation tree of period-1 motion to chaos of the Brusselator is presented through frequency-amplitude characteristics, which be measured in frequency domains. Two main results presented in this book are:

- analytical routes of periodical evolutions to chaos and

- independent period-$(2l + 1)$ evolution to chaos.

This book gives a better understanding of periodic evolutions to chaos in the slow-fast varying Brusselator system, and the bifurcation tree of period-1 evolution to chaos is clearly demonstrated, which can help one understand routes of periodic evolutions to chaos in chemical reaction oscillators. The slow-fast varying systems extensively exist in biological systems and disease dynamical systems. The methodology presented in this book can be used to investigate the slow-fast varying oscillating motions in biological systems and disease dynamical systems for a better understanding of how infectious diseases spread.

KEYWORDS

Brusselator, period-m evolutions, analytical solutions, stability, bifurcations, chaos

傾盆斜佈風

弓驟大樹彎

弓如鈎生北況

浮草低頭故待

明日照九州朝後

Contents

Preface

In 1968, Prigogine and Lefever proposed the Brusselator for a combination of four chemical reactions. Such a Brusselator was extensively studied because of the slow and fast oscillation. In 1971, Lefever and Nicolis proved that a unique limit cycle existed near the unstable equilibrium. Thus, such a limit cycle observed numerically simulated and subharmonic evolutions of chemical concentrations in a coupled Brusselator were observed. For further understanding of slow-fast evolutions in the Brusselator, one considered the periodically forced Brusselator, and the perturbation method and numerical methods were employed for the periodic evolutions of concentration and the jump phenomena. Until now, analytical periodic evolutions to chaos had not been provided.

This book has six chapters. Chapter 1 gives a brief introduction of the Brusselator. The analytical method for periodic flows in nonlinear dynamical systems is presented in Chapter 2. The analytical method is based on a transformation of infinite Fourier series. Such a transformation converted dynamical systems to a system of coefficients in the Fourier series, similar to the Laplace transform of linear dynamical systems. Based on the analytical method, the analytical solutions of periodic evolutions in the periodically diffused Brusselators are discussed in Chapter 3. In Chapter 4, the bifurcation trees of period-1 evolution to chaos is presented through period-1 to period-8 evolutions. The independent period-$(2l + 1)$ evolutions ($l = 1, 2, 3, \cdots$) to chaos are preented in Chapter 5 through period-3,5,7,9 evolutions. In Chapter 6, productions and compensation of catalysts in the entire chemical reactions are discussed. This book provides an analytical route to chaotic evolutions with the slow-fast chaotic oscillations in the periodically diffused Brusselators.

Finally, the authors hope the materials presented herein can give some inspiration on the slow-fast oscillation nonlinear systems in nonlinear dynamics community. The method presented in this book can be used to solve the epidemic diseases spread mathematically. This book is dedicated to all scientists, medical doctors, and nurses fighting Covid-19 for our society.

Albert C.J. Luo and Siyu Guo
Edwardsville, Illinois
May 2020

CHAPTER 1

Introduction

In 1968, the Brusselator for a combination of four chemical reactions was first proposed by Prigogine and Lefever [1]. Such a Brusselator was extensively investigated because of the slow and fast oscillation. In 1971, Lefever and Nicolis [2] proved that a unique limit cycle existed near the unstable equilibrium. Tyson [3] gave an illustration of such a limit cycle and observed subharmonic evolutions of chemical concentrations in a coupled Brusselator. However, in such studies, the diffusion effect was neglected and numerical methods were employed. Thus, no significant achievements were presented.

Tomita et al. [4] investigated the periodically forced Brusselator, and the perturbation method was employed for determining the periodic evolutions of concentration and the jump phenomena. Hao and Zhang [5] adopted a numerical algorithm with double accuracy for the bifurcation diagram of the periodically forced Brusselator. Recently, one used the perturbation method for periodic motions and chaos in nonlinear systems. Choudhury and Tanriver [6] used perturbation method to investigate the homoclinic bifurcation following Hopf bifurcation in nonlinear dynamical systems. Maaita [7] investigated the bifurcation of the slow invariant manifold of completed oscillators. Yamgoue et al. [8] studied the approximate analytical solutions of a constrained nonlinear mechanical system. Shayak and Vyas [9] applied the Krylov-Bogoliubov method to the Mathieu equation. Rajamani and Rajasekar [10] discussed the response amplitude of the parametric Duffing oscillator. Even though chaotic motions in nonlinear systems were investigated through the perturbation analysis, such a way for chaotic motions in nonlinear dynamical systems is inadequate. This is because the perturbation methods required the corresponding linear solutions to determine the approximate periodic solutions of the original nonlinear systems, and the perturbation expansion with small parameters was adopted. Cochelin and Vergez [11] used the harmonic balance method by transferring any nonlinear term higher than a quadratic polynomial to quadratic terms by introducing intermediate variables. However, such a standard process employed the asymptotic numerical method, which was not accurate to calculate the harmonic coefficients.

To improve the traditional harmonic balance method, Luo [12] proposed the generalized harmonic balance method. Such a method was based on a transformation of infinite Fourier series, which converted dynamical systems to a system of harmonic coefficients in Fourier series. In 2012, Luo and Huang [13, 14] applied such a generalized harmonic balance method to a Duffing oscillator. Bifurcation trees of periodic motions to chaos were presented for a better understanding of the nonlinear characteristics of periodic motions. The analytical bifurcation trees of periodic motions in the Duffing oscillator to chaos were obtained (see also Luo and Huang [15–20]). Such analytical bifurcation trees showed the connection from periodic motions to chaos analytically. Luo and Yu [21] used the generalized harmonic balance method for approximate analytical solutions of period-1 motions in a nonlinear quadratic oscillator. Luo and Lakeh [22] presented the analytical solutions of periodic motions in the van der Pol oscillator through the generalized harmonic balance method. Luo and Lakeh [23] obtained the analytical solutions of period-m motions of the van der Pol-duffing oscillator. Luo and Lakeh [24] presented the bifurcation trees of periodic motion to chaos in the van der Pol-duffing oscillator. In 2016, Luo and Yu [25, 26] discussed the analytical solutions for the bifurcation trees of period-1 motions to chaos in a two-degree-of-freedom nonlinear oscillator. In 2013, Luo [27] extended such ideas of the generalized harmonic method for periodic motions in time-delay, nonlinear dynamical systems. Luo and Jin [28] applied such a method for the bifurcation tree of period-1 motion to chaos in a periodically forced, quadratic nonlinear oscillator with time-delay. Further, Luo and Jin [29–31] investigated the periodic motions to chaos in a time-delayed Duffing oscillator. The generalized harmonic balance method is suitable for periodic motions in nonlinear systems with polynomial nonlinearity. From the generalized harmonic balance method, Wang and Liu [32] is a numerical scheme to compute coefficients in the finite Fourier series expression of periodic motions in nonlinear dynamical systems. Luo and Wang [33] used the generalized harmonic balance method for approximate analytical solutions of periodic motions in the rotor dynamical systems. Akhmet and Fen [34] investigated almost periodicity of chaos. The generalized harmonic balance methods and applications can be found in Luo [35, 36]. In Luo and Guo [37, 38], the bifurcation trees of period-1 evolutions to chaos in the periodically forced Brusselator were studied. However, for such slow-fast varying oscillator, there are many independent periodic evolutions, as presented in Luo and Guo [39].

To help one understand the slow-fast oscillating autocatalytic reactions, the analytical periodic motions to chaos in the periodically diffused Brusselators is presented in this book. Chapter 1 gives a brief introduction of the Brusselator. In Chapter 2, the analytical method for periodic flows in nonlinear dynamical systems is presented. Based on the analytical method, the analytical solutions of periodic evolutions in the periodically diffused Brusselators will be discussed in Chapter 3. In Chapter 4, the bifurcation trees of period-1 evolution to chaos will be presented through period-1 to period-8 evolutions. The independent period-$(2l + 1)$ evolutions ($l = 1, 2, 3, \ldots$) to chaos will be presented in Chapter 5 through period-3, 5, 7, 9 evolutions. In Chapter 6, productions and compensation of catalysts in the entire chemical reactions will be discussed.

Generalized Harmonic Balance Method

In this chapter, from Luo [12], the analytical dynamics of periodic flows and chaos in nonlinear dynamical systems is presented. The analytical solutions of periodic flows and chaos in autonomous systems will be discussed first, and then the analytical dynamics of non-autonomous nonlinear dynamical systems will be presented. An alternative way for the analytical solution is presented in Luo [35, 36].

2.1   AUTONOMOUS NONLINEAR SYSTEMS

Periodic flows in autonomous dynamical systems will be presented herein. If an autonomous nonlinear system has a periodic flow with period $T = 2\pi/\Omega$, then such a periodic flow can be expressed by a generalized coordinate transformation based on the Fourier series. The method is stated through the following theorem.

Theorem 2.1   *Consider a nonlinear dynamical system as*

$$\dot{\mathbf{x}} = \mathbf{f}(\mathbf{x}, \mathbf{p}) \in \Re^n, \tag{2.1}$$

where $\mathbf{f}(\mathbf{x}, \mathbf{p})$ is a C^r-continuous nonlinear function vector ($r \geq 1$). If such a dynamical system has a periodic flow $\mathbf{x}(t)$ with finite norm $\|\mathbf{x}\|$ and period $T = 2\pi/\Omega$, there is a generalized coordinate transformation with $\theta = \Omega t$ for the periodic flow of Eq. (2.1) in a form of

$$\mathbf{x}(t) = \mathbf{a}_0(t) + \sum_{k=1}^{\infty} \mathbf{b}_k(t)\cos(k\theta) + \mathbf{c}_k(t)\sin(k\theta) \tag{2.2}$$

with

$$
\begin{aligned}
\mathbf{a}_0 &= (a_{01}, a_{02}, \ldots, a_{0n})^{\mathrm{T}}, \\
\mathbf{b}_k &= (b_{k1}, b_{k2}, \ldots, b_{kn})^{\mathrm{T}}, \\
\mathbf{c}_k &= (c_{k1}, c_{k2}, \ldots, c_{kn})^{\mathrm{T}}
\end{aligned}
\tag{2.3}
$$

and

$$\|\mathbf{x}\| = \|\mathbf{a}_0\| + \sum_{k=1}^{\infty} \|\mathbf{A}_k\|, \quad \text{and} \quad \lim_{k \to \infty} \|\mathbf{A}_k\| = 0 \quad \text{but not uniform}$$

$$\text{with} \quad \mathbf{A}_k = (A_{k1}, A_{k2}, \dots, A_{kn})^{\mathrm{T}} \quad \text{and} \quad A_{kj} = \sqrt{b_{kj}^2 + c_{kj}^2} \quad (j = 1, 2, \dots, n).$$

(2.4)

For $\|\mathbf{x}(t) - \mathbf{x}^*(t)\| < \varepsilon$ with a prescribed small positive $\varepsilon > 0$, the infinite term transformation of the periodic flow $\mathbf{x}(t)$ of Eq. (2.1), given by Eq. (2.2), can be approximated by a finite term transformation $\mathbf{x}^*(t)$ as

$$\mathbf{x}^*(t) = \mathbf{a}_0(t) + \sum_{k=1}^{N} \mathbf{b}_k(t) \cos(k\theta) + \mathbf{c}_k(t) \sin(k\theta) \tag{2.5}$$

and the generalized coordinates are determined by

$$\dot{\mathbf{a}}_0 = \mathbf{F}_0\left(\mathbf{a}_0, \mathbf{b}, \mathbf{c}\right),$$
$$\dot{\mathbf{b}} = -\Omega \mathbf{k}_1 \mathbf{c} + \mathbf{F}_1\left(\mathbf{a}_0, \mathbf{b}, \mathbf{c}\right), \tag{2.6}$$
$$\dot{\mathbf{c}} = \Omega \mathbf{k}_1 \mathbf{b} + \mathbf{F}_2\left(\mathbf{a}_0, \mathbf{b}, \mathbf{c}\right),$$

where

$$\mathbf{k}_1 = diag\left(\mathbf{I}_{n \times n}, 2\mathbf{I}_{n \times n}, \dots, N\mathbf{I}_{n \times n}\right),$$
$$\mathbf{b} = (\mathbf{b}_1, \mathbf{b}_2, \dots, \mathbf{b}_N)^{\mathrm{T}},$$
$$\mathbf{c} = (\mathbf{c}_1, \mathbf{c}_2, \dots, \mathbf{c}_N)^{\mathrm{T}},$$
$$\mathbf{F}_1 = (\mathbf{F}_{11}, \mathbf{F}_{12}, \dots, \mathbf{F}_{1N})^{\mathrm{T}}, \tag{2.7}$$
$$\mathbf{F}_2 = (\mathbf{F}_{21}, \mathbf{F}_{22}, \dots, \mathbf{F}_{2N})^{\mathrm{T}}$$
$$\text{for} \quad N = 1, 2, \dots, \infty$$

and

$$\mathbf{F}_0\left(\mathbf{a}_0, \mathbf{b}, \mathbf{c}\right) = \frac{1}{2\pi} \int_0^{2\pi} \mathbf{f}(\mathbf{x}^*, \mathbf{p}) d\theta;$$
$$\mathbf{F}_{1k}\left(\mathbf{a}_0, \mathbf{b}, \mathbf{c}\right) = \frac{1}{\pi} \int_0^{2\pi} \mathbf{f}(\mathbf{x}^*, \mathbf{p}) \cos(k\theta) d\theta, \tag{2.8}$$
$$\mathbf{F}_{2k}\left(\mathbf{a}_0, \mathbf{b}, \mathbf{c}\right) = \frac{1}{\pi} \int_0^{2\pi} \mathbf{f}(\mathbf{x}^*, \mathbf{p}) \sin(k\theta) d\theta$$
$$\text{for} \quad k = 1, 2, \dots, N.$$

Equation (2.6) becomes

$$\dot{\mathbf{z}} = \mathbf{f}(\mathbf{z}), \tag{2.9}$$

where

$$\mathbf{z} = (\mathbf{a}_0, \mathbf{b}, \mathbf{c})^{\mathrm{T}},$$
$$\mathbf{f} = (\mathbf{F}_0, -\Omega \mathbf{k}_1 \mathbf{c} + \mathbf{F}_1, \Omega \mathbf{k}_1 \mathbf{b} + \mathbf{F}_2)^{\mathrm{T}}. \tag{2.10}$$

If equilibrium $\mathbf{z}^*$ of Eq. (2.6) (i.e., $\mathbf{f}(\mathbf{z}^*) = 0$) exists, then the approximate solution of periodic flow exists as in Eq. (2.5). In vicinity of equilibrium $\mathbf{z}^*$, with $\mathbf{z} = \mathbf{z}^* + \Delta \mathbf{z}$, the linearized equation of Eq. (2.4) is

$$\Delta \dot{\mathbf{z}} = D\mathbf{f}(\mathbf{z}^*)\Delta \mathbf{z} \tag{2.11}$$

and the eigenvalue analysis of equilibrium $\mathbf{z}^*$ is given by

$$\left| D\mathbf{f}(\mathbf{z}^*) - \lambda \mathbf{I}_{n(2N+1) \times n(2N+1)} \right| = 0, \tag{2.12}$$

where $D\mathbf{f}(\mathbf{z}^*) = \partial \mathbf{f}(\mathbf{z})/\partial(\mathbf{z})|_{\mathbf{z}^*}$. Thus, the stability and bifurcation of periodic flow can be classified by the eigenvalues of $D\mathbf{f}(\mathbf{z}^*)$ with

$$(n_1, n_2, n_3 | n_4, n_5, n_6), \tag{2.13}$$

where n_1 is the total number of negative real eigenvalues, n_2 is the total number of positive real eigenvalues, n_3 is the total number of zero real eigenvalues, n_4 is the total pair number of complex eigenvalues with negative real parts, n_5 is the total pair number of complex eigenvalues with positive real parts, and n_6 is the total pair number of complex eigenvalues with zero real parts.

(i) If all eigenvalues of the equilibrium possess negative real parts, the approximate periodic solution is stable.

(ii) If at least one of eigenvalues of the equilibrium possesses positive real part, the approximate periodic solution is unstable.

(iii) The boundaries between stable and unstable equilibriums with higher-order singularity give bifurcation and stability conditions with higher-order singularity.

Proof. See Luo [12, 35, 36]. □

If the Hopf bifurcation of equilibriums of Eq. (2.6) occurs, a periodic solution of the generalized coordinates in Eq. (2.5) exists with a frequency ω. Thus, a periodic solution of the generalized coordinates is given through a new generalized coordinate transformation with $\vartheta = \omega t$

$$\mathbf{a}_0(t) = \mathbf{A}_{00}(t) + \sum_{m=1}^{\infty} \mathbf{A}_{0m}^{(1)}(t)\cos(m\vartheta) + \mathbf{A}_{0m}^{(2)}(t)\sin(m\vartheta),$$

$$\mathbf{b}_k(t) = \mathbf{B}_{k0}(t) + \sum_{m=1}^{\infty} \mathbf{B}_{km}^{(1)}(t)\cos(m\vartheta) + \mathbf{B}_{km}^{(2)}(t)\sin(m\vartheta), \qquad (2.14)$$

$$\mathbf{c}_k(t) = \mathbf{C}_{k0}(t) + \sum_{m=1}^{\infty} \mathbf{C}_{km}^{(1)}(t)\cos(m\vartheta) + \mathbf{C}_{km}^{(2)}(t)\sin(m\vartheta).$$

Substitution of Eq. (2.14) into Eq. (2.5) yields

$$
\begin{aligned}
\mathbf{x}(t) = &\, \mathbf{A}_{00}(t) + \sum_{m=1}^{\infty} \mathbf{A}_{0m}^{(1)}(t)\cos(m\vartheta) + \mathbf{A}_{0m}^{(2)}(t)\sin(m\vartheta) \\
&+ \sum_{k=1}^{\infty} \mathbf{B}_{k0}(t)\cos(k\theta) + \mathbf{C}_{k0}(t)\sin(k\theta) \\
&+ \sum_{k=1}^{\infty}\sum_{m=1}^{\infty} \mathbf{B}_{km}^{(1)}(t)\cos(m\vartheta)\cos(k\theta) + \mathbf{B}_{km}^{(2)}(t)\sin(m\vartheta)\cos(k\theta) \\
&+ \sum_{k=1}^{\infty}\sum_{m=1}^{\infty} \mathbf{C}_{km}^{(1)}(t)\cos(m\vartheta)\sin(k\theta) + \mathbf{C}_{km}^{(2)}(t)\sin(m\vartheta)\sin(k\theta).
\end{aligned}
\qquad (2.15)
$$

If a new solution is still periodic with excitation period $T = 2\pi/\Omega$, then for specific m, the following relation exists:

$$m\vartheta = k\theta \quad \Rightarrow \quad m\omega = k\Omega. \qquad (2.16)$$

For this case, $k = 1$ should be employed because $k > 1$ terms are included in the Fourier series expression. Thus,

$$m\vartheta = \theta \quad \Rightarrow \quad m\omega = \Omega. \qquad (2.17)$$

For $m = 1$, the period-1 flow is obtained, and Eq. (2.15) becomes Eq. (2.5). For the period-m flow, we have a new generalized coordinate transformation as

$$\mathbf{x}^{(m)}(t) = \mathbf{a}_0^{(m)}(t) + \sum_{k=1}^{\infty} \mathbf{b}_{k/m}(t)\cos\left(\frac{k}{m}\theta\right) + \mathbf{c}_{k/m}(t)\sin\left(\frac{k}{m}\theta\right). \qquad (2.18)$$

If $||\mathbf{x}^{(m)}(t) - \mathbf{x}^{(m)*}(t)|| < \varepsilon$ with a prescribed small $\varepsilon > 0$, the solution of period-m flow in Eq. (2.18) can be approximated by a finite term transformation as

$$\mathbf{x}^{(m)*}(t) = \mathbf{a}_0^{(m)}(t) + \sum_{k=1}^{N} \mathbf{b}_{k/m}(t)\cos\left(\frac{k}{m}\theta\right) + \mathbf{c}_{k/m}(t)\sin\left(\frac{k}{m}\theta\right). \qquad (2.19)$$

If $m\omega \neq k\Omega$ for any m and k, the solution is quasi-periodic or chaotic instead of periodic. If period-1 motion possesses at least N_1 harmonic vector terms, then the total harmonic vector terms for period-m motion should be $N \geq mN_1$. Similarly, period-m flows in nonlinear dynamical systems can be discussed.

Theorem 2.2   *Consider a nonlinear dynamical system in Eq. (2.1). If such a dynamical system has a period-m flow $\mathbf{x}^{(m)}(t)$ with finite norm $||\mathbf{x}^{(m)}||$ and period $T = 2\pi/\Omega$, there is a generalized coordinate transformation with $\theta = \Omega t$ for the period-m flow of Eq. (2.1) in the form of*

$$\mathbf{x}^{(m)}(t) = \mathbf{a}_0^{(m)}(t) + \sum_{k=1}^{\infty} \mathbf{b}_{k/m}(t) \cos\left(\frac{k}{m}\theta\right) + \mathbf{c}_{k/m}(t) \sin\left(\frac{k}{m}\theta\right) \tag{2.20}$$

with

$$\begin{aligned}
\mathbf{a}_0^{(m)} &= \left(a_{01}^{(m)}, a_{02}^{(m)}, \ldots, a_{0n}^{(m)}\right)^{\mathrm{T}}, \\
\mathbf{b}_{k/m} &= \left(b_{k/m1}, b_{k/m2}, \ldots, b_{k/mn}\right)^{\mathrm{T}}, \\
\mathbf{c}_{k/m} &= \left(c_{k/m1}, c_{k/m2}, \ldots, c_{k/mn}\right)^{\mathrm{T}}
\end{aligned} \tag{2.21}$$

and

$$||\mathbf{x}^{(m)}|| = ||\mathbf{a}_0^{(m)}|| + \sum_{k=1}^{\infty} ||\mathbf{A}_{k/m}||, \quad \text{and} \quad \lim_{k \to \infty} ||\mathbf{A}_{k/m}|| = 0 \quad \text{but not uniform}$$

with $\mathbf{A}_{k/m} = \left(A_{k/m1}, A_{k/m2}, \ldots, A_{k/mn}\right)^{\mathrm{T}}$

and $A_{k/mj} = \sqrt{b_{k/mj}^2 + c_{k/mj}^2}$ $(j = 1, 2, \ldots, n).$

$$\tag{2.22}$$

For $||\mathbf{x}^{(m)}(t) - \mathbf{x}^{(m)*}(t)|| < \varepsilon$ with a prescribed small $\varepsilon > 0$, the infinite term transformation $\mathbf{x}^{(m)}(t)$ of the period-m flow of Eq. (2.1), given by Eq. (2.20), can be approximated by a finite term transformation $\mathbf{x}^{(m)*}(t)$ as

$$\mathbf{x}^{(m)*}(t) = \mathbf{a}_0^{(m)}(t) + \sum_{k=1}^{N} \mathbf{b}_{k/m}(t) \cos\left(\frac{k}{m}\theta\right) + \mathbf{c}_{k/m}(t) \sin\left(\frac{k}{m}\theta\right) \tag{2.23}$$

and the generalized coordinates are determined by

$$\begin{aligned}
\dot{\mathbf{a}}_0^{(m)} &= \mathbf{F}_0^{(m)}(\mathbf{a}_0^{(m)}, \mathbf{b}^{(m)}, \mathbf{c}^{(m)}), \\
\dot{\mathbf{b}}^{(m)} &= -\frac{\Omega}{m}\mathbf{k}_1\mathbf{c}^{(m)} + \mathbf{F}_1^{(m)}(\mathbf{a}_0^{(m)}, \mathbf{b}^{(m)}, \mathbf{c}^{(m)}), \\
\dot{\mathbf{c}}^{(m)} &= \frac{\Omega}{m}\mathbf{k}_1\mathbf{b}^{(m)} + \mathbf{F}_2^{(m)}(\mathbf{a}_0^{(m)}, \mathbf{b}^{(m)}, \mathbf{c}^{(m)}),
\end{aligned} \tag{2.24}$$

where

$$k_1 = diag\left(\mathbf{I}_{n \times n}, 2\mathbf{I}_{n \times n}, \ldots, N\mathbf{I}_{n \times n}\right),$$
$$\mathbf{b}^{(m)} = \left(\mathbf{b}_{1/m}, \mathbf{b}_{2/m}, \ldots, \mathbf{b}_{N/m}\right)^{\mathrm{T}},$$
$$\mathbf{c}^{(m)} = \left(\mathbf{c}_{1/m}, \mathbf{c}_{2/m}, \ldots, \mathbf{c}_{N/m}\right)^{\mathrm{T}},$$
$$\mathbf{F}_1^{(m)} = \left(\mathbf{F}_{11}^{(m)}, \mathbf{F}_{12}^{(m)}, \ldots, \mathbf{F}_{1N}^{(m)}\right)^{\mathrm{T}}, \tag{2.25}$$
$$\mathbf{F}_2^{(m)} = \left(\mathbf{F}_{21}^{(m)}, \mathbf{F}_{22}^{(m)}, \ldots, \mathbf{F}_{2N}^{(m)}\right)^{\mathrm{T}}$$

for $N = 1, 2, \ldots, \infty;$

and

$$\mathbf{F}_0^{(m)}(\mathbf{a}_0^{(m)}, \mathbf{b}^{(m)}, \mathbf{c}^{(m)}) = \frac{1}{2m\pi} \int_0^{2m\pi} \mathbf{f}(\mathbf{x}^{(m)*}, \mathbf{p}) d\theta,$$
$$\mathbf{F}_{1k}^{(m)}(\mathbf{a}_0^{(m)}, \mathbf{b}^{(m)}, \mathbf{c}^{(m)}) = \frac{1}{m\pi} \int_0^{2m\pi} \mathbf{f}(\mathbf{x}^{(m)*}, \mathbf{p}) \cos\left(\frac{k}{m}\theta\right) d\theta, \tag{2.26}$$
$$\mathbf{F}_{2k}^{(m)}(\mathbf{a}_0^{(m)}, \mathbf{b}^{(m)}, \mathbf{c}^{(m)}) = \frac{1}{m\pi} \int_0^{2m\pi} \mathbf{f}(\mathbf{x}^{(m)*}, \mathbf{p}) \sin\left(\frac{k}{m}\theta\right) d\theta$$

for $k = 1, 2, \ldots, N.$

Equation (2.24) *becomes*

$$\dot{\mathbf{z}}^{(m)} = \mathbf{f}^{(m)}(\mathbf{z}^{(m)}) \tag{2.27}$$

where

$$\mathbf{z}^{(m)} = \left(\mathbf{a}_0^{(m)}, \mathbf{b}^{(m)}, \mathbf{c}^{(m)}\right)^{\mathrm{T}}$$
$$\mathbf{f}^{(m)} = \left(\mathbf{F}_0^{(m)}, -\frac{1}{m}\Omega k_1 \mathbf{c}^{(m)} + \mathbf{F}_1^{(m)}, \frac{1}{m}\Omega k_1 \mathbf{b}^{(m)} + \mathbf{F}_2^{(m)}\right)^{\mathrm{T}}. \tag{2.28}$$

If equilibrium $\mathbf{z}^{(m)*}$ *of Eq.* (2.27) *(i.e.,* $\mathbf{f}^{(m)}(\mathbf{z}^{(m)*}) = \mathbf{0}$ *) exists, then the approximate solution of period-m flow exists as in Eq.* (2.23). *In vicinity of equilibrium* $\mathbf{z}^{(m)*}$, *with* $\mathbf{z}^{(m)} = \mathbf{z}^{(m)*} + \Delta\mathbf{z}^{(m)}$, *the linearized equation of Eq.* (2.27) *is*

$$\Delta\dot{\mathbf{z}}^{(m)} = D\mathbf{f}^{(m)}(\mathbf{z}^{(m)*})\Delta\mathbf{z}^{(m)} \tag{2.29}$$

and the eigenvalue analysis of equilibrium $\mathbf{z}^*$ *is given by*

$$\left| D\mathbf{f}^{(m)}(\mathbf{z}^{(m)*}) - \lambda\mathbf{I}_{n(2N+1) \times n(2N+1)} \right| = 0, \tag{2.30}$$

where $D\mathbf{f}^{(m)}(\mathbf{z}^{(m)*}) = \partial\mathbf{f}^{(m)}(\mathbf{z}^{(m)})/\partial\mathbf{z}^{(m)}\big|_{\mathbf{z}^{(m)*}}$. *The stability and bifurcation of such a periodic solution can be classified by the eigenvalues of* $D\mathbf{f}^{(m)}(\mathbf{z}^{(m)*})$ *with*

$$(n_1, n_2, n_3 | n_4, n_5, n_6),\qquad(2.31)$$

where n_1 *is the total number of negative real eigenvalues,* n_2 *is the total number of positive real eigenvalues,* n_3 *is the total number of zero real eigenvalues,* n_4 *is the total pair number of complex eigenvalues with negative real parts,* n_5 *is the total pair number of complex eigenvalues with positive real parts, and* n_6 *is the total pair number of complex eigenvalues with zero real parts.*

(i) *If all eigenvalues of the equilibrium possess negative real parts, the approximate periodic solution is stable.*

(ii) *If at least one of eigenvalues of the equilibrium possesses positive real part, the approximate periodic solution is unstable.*

(iii) *The boundaries between stable and unstable equilibriums with higher-order singularity give bifurcation and stability conditions with higher-order singularity.*

Proof. See Luo [35, 36]. □

2.2   NON-AUTONOMOUS NONLINEAR SYSTEMS

If a periodically excited nonlinear system with an external period $T = 2\pi/\Omega$ has a periodic flow, then such a periodic flow can be expressed by a Fourier series. As in Luo [35, 36], the following theorems are presented.

Theorem 2.3   *Consider a non-autonomous nonlinear dynamical system as*

$$\dot{\mathbf{x}} = \mathbf{F}(\mathbf{x}, t, \mathbf{p}) \in \Re^n,\qquad(2.32)$$

where $\mathbf{F}(\mathbf{x}, t, \mathbf{p})$ *is a* C^r*-continuous nonlinear function vector* $(r \geq 1)$ *with an excitation period* $T = 2\pi/\Omega$. *If such a dynamical system has a periodic flow* $\mathbf{x}(t)$ *with finite norm* $||\mathbf{x}||$, *there is a generalized coordinate transformation with* $\theta = \Omega t$ *for the periodic flow of Eq.* (2.32) *in a form of*

$$\mathbf{x}(t) = \mathbf{a}_0(t) + \sum_{k=1}^{\infty} \mathbf{b}_k(t)\cos(k\theta) + \mathbf{c}_k(t)\sin(k\theta)\qquad(2.33)$$

with

$$\begin{aligned}
\mathbf{a}_0 &= (a_{01}, a_{02}, \ldots, a_{0n})^{\mathrm{T}},\\
\mathbf{b}_k &= (b_{k1}, b_{k2}, \ldots, b_{kn})^{\mathrm{T}},\\
\mathbf{c}_k &= (c_{k1}, c_{k2}, \ldots, c_{kn})^{\mathrm{T}};
\end{aligned}\qquad(2.34)$$

and

$$||\mathbf{x}|| = ||\mathbf{a}_0|| + \sum_{k=1}^{\infty} ||\mathbf{A}_k||, \quad \text{and} \quad \lim_{k\to\infty} ||\mathbf{A}_k|| = 0 \quad \text{but not uniform} \tag{2.35}$$

with $\mathbf{A}_k = (A_{k1}, A_{k2}, \ldots, A_{kn})^{\mathrm{T}}$ and $A_{kj} = \sqrt{b_{kj}^2 + c_{kj}^2}$ $(j = 1, 2, \ldots, n)$.

For $||\mathbf{x}(t) - \mathbf{x}^*(t)|| < \varepsilon$ with a prescribed small positive $\varepsilon > 0$, the infinite term transformation of the periodic flow $\mathbf{x}(t)$ of Eq. (2.32), given by Eq. (2.33), can be approximated by a finite term transformation $\mathbf{x}^*(t)$ as

$$\mathbf{x}^*(t) = \mathbf{a}_0(t) + \sum_{k=1}^{N} \mathbf{b}_k(t) \cos(k\theta) + \mathbf{c}_k(t) \sin(k\theta) \tag{2.36}$$

and the generalized coordinates are determined by

$$\begin{aligned} \dot{\mathbf{a}}_0 &= \mathbf{F}_0(\mathbf{a}_0, \mathbf{b}, \mathbf{c}), \\ \dot{\mathbf{b}} &= -\Omega \mathbf{k}_1 \mathbf{c} + \mathbf{F}_1(\mathbf{a}_0, \mathbf{b}, \mathbf{c}), \\ \dot{\mathbf{c}} &= \Omega \mathbf{k}_1 \mathbf{b} + \mathbf{F}_2(\mathbf{a}_0, \mathbf{b}, \mathbf{c}); \end{aligned} \tag{2.37}$$

where

$$\begin{aligned} \mathbf{k}_1 &= diag(\mathbf{I}_{n\times n}, 2\mathbf{I}_{n\times n}, \ldots, N\mathbf{I}_{n\times n}), \\ \mathbf{b} &= (\mathbf{b}_1, \mathbf{b}_2, \ldots, \mathbf{b}_N)^{\mathrm{T}}, \\ \mathbf{c} &= (\mathbf{c}_1, \mathbf{c}_2, \ldots, \mathbf{c}_N)^{\mathrm{T}}, \\ \mathbf{F}_1 &= (\mathbf{F}_{11}, \mathbf{F}_{12}, \ldots, \mathbf{F}_{1N})^{\mathrm{T}}, \\ \mathbf{F}_2 &= (\mathbf{F}_{21}, \mathbf{F}_{22}, \ldots, \mathbf{F}_{2N})^{\mathrm{T}}, \\ \text{for} \quad N &= 1, 2, \ldots, \infty; \end{aligned} \tag{2.38}$$

and for $k = 1, 2, \ldots, N$

$$\begin{aligned} \mathbf{F}_0(\mathbf{a}_0, \mathbf{b}, \mathbf{c}) &= \frac{1}{2\pi} \int_0^{2\pi} \mathbf{F}(\mathbf{x}^*, t, \mathbf{p}) d\theta, \\ \mathbf{F}_{1k}(\mathbf{a}_0, \mathbf{b}, \mathbf{c}) &= \frac{1}{\pi} \int_0^{2\pi} \mathbf{F}(\mathbf{x}^*, t, \mathbf{p}) \cos(k\theta) d\theta, \\ \mathbf{F}_{2k}(\mathbf{a}_0, \mathbf{b}, \mathbf{c}) &= \frac{1}{\pi} \int_0^{2\pi} \mathbf{F}(\mathbf{x}^*, t, \mathbf{p}) \sin(k\theta) d\theta. \end{aligned} \tag{2.39}$$

Equation (2.37) *becomes*

$$\dot{\mathbf{z}} = \mathbf{f}(\mathbf{z}) \tag{2.40}$$

where

$$z = (a_0, b, c)^{\mathrm{T}},$$
$$f = (F_0, -\Omega k_1 c + F_1, \Omega k_1 b + F_2)^{\mathrm{T}}. \tag{2.41}$$

If equilibrium z^* of Eq. (2.40) (i.e., $f(z^*) = 0$) exists, then the approximate solution of periodic flow exists as in Eq. (2.36). In vicinity of equilibrium z^*, with $z = z^* + \Delta z$ the linearized equation of Eq. (2.40) is

$$\Delta \dot{z} = Df(z^*)\Delta z \tag{2.42}$$

and the eigenvalue analysis of equilibrium z^* is given by

$$\left| Df(z^*) - \lambda I_{n(2N+1) \times n(2N+1)} \right| = 0, \tag{2.43}$$

where $Df(z^*) = \partial f(z)/\partial z|_{z^*}$. Thus, the stable and bifurcation of periodic solution can be classified by the eigenvaluses of $Df(z^*)$ with

$$(n_1, n_2, n_3 | n_4, n_5, n_6). \tag{2.44}$$

(i) If all eigenvalues of the equilibrium possess negative real parts, the approximate periodic solution is stable.

(ii) If at least one of eigenvalues of the equilibrium possesses positive real part, the approximate periodic solution is unstable.

(iii) The boundaries between stable and unstable equilibriums with higher-order singularity give bifurcation and stability conditions with higher-order singularity.

Proof. The proof can be referred to Luo [12, 35, 36].   □

As similar to the autonomous nonlinear system, period-*m* flows in the periodically excited, nonlinear dynamical system in Eq. (2.32) can be discussed.

Theorem 2.4   *Consider an autonomous nonlinear dynamical system in Eq. (2.32) with an excitation period $T = 2\pi/\Omega$. If such a dynamical system has a period-m flow $x^{(m)}(t)$ with finite norm $||x^{(m)}||$, there is a generalized coordinate transformation with $\theta = \Omega t$ for the periodic flow of Eq. (2.32) in the form of*

$$x^{(m)}(t) = a_0^{(m)}(t) + \sum_{k=1}^{\infty} b_{k/m}(t) \cos\left(\frac{k}{m}\theta\right) + c_{k/m}(t) \sin\left(\frac{k}{m}\theta\right) \tag{2.45}$$

with

$$\mathbf{a}_0^{(m)} = (a_{01}^{(m)}, a_{02}^{(m)}, \ldots, a_{0n}^{(m)})^{\mathrm{T}},$$
$$\mathbf{b}_{k/m} = (b_{k/m1}, b_{k/m2}, \ldots, b_{k/mn})^{\mathrm{T}}, \tag{2.46}$$
$$\mathbf{c}_{k/m} = (c_{k/m1}, c_{k/m2}, \ldots, c_{k/mn})^{\mathrm{T}},$$

and

$$\|\mathbf{x}^{(m)}\| = \|\mathbf{a}_0^{(m)}\| + \sum_{k=1}^{\infty} \|\mathbf{A}_{k/m}\|, \quad \text{and} \quad \lim_{k \to \infty} \|\mathbf{A}_{k/m}\| = 0 \quad \text{but not uniform}$$

with $\mathbf{A}_{k/m} = (A_{k/m1}, A_{k/m2}, \ldots, A_{k/mn})^{\mathrm{T}}$

and $A_{k/mj} = \sqrt{b_{k/mj}^2 + c_{k/mj}^2} \quad (j = 1, 2, \ldots, n).$
$$\tag{2.47}$$

For $\|\mathbf{x}^{(m)}(t) - \mathbf{x}^{(m)*}(t)\| < \varepsilon$ with a prescribed small $\varepsilon > 0$, the infinite term transformation $\mathbf{x}^{(m)}(t)$ of the period-m flow of Eq. (2.32), given by Eq. (2.45), can be approximated by a finite term transformation $\mathbf{x}^{(m)*}(t)$ as

$$\mathbf{x}^*(t) = \mathbf{a}_0^{(m)}(t) + \sum_{k=1}^{N} \mathbf{b}_{k/m}(t) \cos\left(\frac{k}{m}\theta\right) + \mathbf{c}_{k/m}(t) \sin\left(\frac{k}{m}\theta\right) \tag{2.48}$$

and the generalized coordinates are determined by

$$\dot{\mathbf{a}}_0^{(m)} = \mathbf{F}_0^{(m)}(\mathbf{a}_0^{(m)}, \mathbf{b}^{(m)}, \mathbf{c}^{(m)}),$$
$$\dot{\mathbf{b}}^{(m)} = -\frac{\Omega}{m} \mathbf{k}_1 \mathbf{c}^{(m)} + \mathbf{F}_1^{(m)}(\mathbf{a}_0^{(m)}, \mathbf{b}^{(m)}, \mathbf{c}^{(m)}), \tag{2.49}$$
$$\dot{\mathbf{c}}^{(m)} = \frac{\Omega}{m} \mathbf{k}_1 \mathbf{b}^{(m)} + \mathbf{F}_2^{(m)}(\mathbf{a}_0^{(m)}, \mathbf{b}^{(m)}, \mathbf{c}^{(m)}),$$

where for $N = 1, 2, \ldots, \infty$

$$\mathbf{k}_1 = diag\left(\mathbf{I}_{n \times n}, 2\mathbf{I}_{n \times n}, \ldots, N\mathbf{I}_{n \times n}\right),$$
$$\mathbf{b}^{(m)} = \left(\mathbf{b}_{1/m}, \mathbf{b}_{2/m}, \ldots, \mathbf{b}_{N/m}\right)^{\mathrm{T}},$$
$$\mathbf{c}^{(m)} = \left(\mathbf{c}_{1/m}, \mathbf{c}_{2/m}, \ldots, \mathbf{c}_{N/m}\right)^{\mathrm{T}}, \tag{2.50}$$
$$\mathbf{F}_1^{(m)} = \left(\mathbf{F}_{11}^{(m)}, \mathbf{F}_{12}^{(m)}, \ldots, \mathbf{F}_{1N}^{(m)}\right)^{\mathrm{T}},$$
$$\mathbf{F}_2^{(m)} = \left(\mathbf{F}_{21}^{(m)}, \mathbf{F}_{22}^{(m)}, \ldots, \mathbf{F}_{2N}^{(m)}\right)^{\mathrm{T}},$$

and

$$\mathbf{F}_0^{(m)}(\mathbf{a}_0^{(m)},\mathbf{b}^{(m)},\mathbf{c}^{(m)}) = \frac{1}{2m\pi}\int_0^{2m\pi}\mathbf{F}(\mathbf{x}^{(m)*},t,\mathbf{p})d\theta,$$

$$\mathbf{F}_{1k}^{(m)}(\mathbf{a}_0^{(m)},\mathbf{b}^{(m)},\mathbf{c}^{(m)}) = \frac{1}{m\pi}\int_0^{2m\pi}\mathbf{F}(\mathbf{x}^{(m)*},t,\mathbf{p})\cos\left(\frac{k}{m}\theta\right)d\theta, \qquad (2.51)$$

$$\mathbf{F}_{2k}^{(m)}(\mathbf{a}_0^{(m)},\mathbf{b}^{(m)},\mathbf{c}^{(m)}) = \frac{1}{m\pi}\int_0^{2m\pi}\mathbf{F}(\mathbf{x}^{(m)*},t,\mathbf{p})\sin\left(\frac{k}{m}\theta\right)d\theta,$$

for $k = 1, 2, \ldots, N$.

Equation (2.49) *becomes*

$$\dot{\mathbf{z}}^{(m)} = \mathbf{f}^{(m)}(\mathbf{z}^{(m)}), \qquad (2.52)$$

where

$$\mathbf{z}^{(m)} = \left(\mathbf{a}_0^{(m)},\mathbf{b}^{(m)},\mathbf{c}^{(m)}\right)^{\mathrm{T}},$$

$$\mathbf{f}^{(m)} = \left(\mathbf{F}_0^{(m)}, -\Omega\mathbf{k}_1\mathbf{c}^{(m)}/m + \mathbf{F}_1^{(m)}, \Omega\mathbf{k}_1\mathbf{b}^{(m)}/m + \mathbf{F}_2^{(m)}\right)^{\mathrm{T}}. \qquad (2.53)$$

If equilibrium $\mathbf{z}^{(m)*}$ *of Eq.* (2.52) (*i.e.,* $\mathbf{f}^{(m)}(\mathbf{z}^{(m)*}) = 0$) *exists, then the approximate solution of period-m flow exists as in Eq.* (2.48). *In vicinity of equilibrium* $\mathbf{z}^{(m)*}$, *with* $\mathbf{z}^{(m)} = \mathbf{z}^{(m)*} + \Delta\mathbf{z}^{(m)}$, *the linearized equation of Eq.* (2.52) *is*

$$\Delta\dot{\mathbf{z}}^{(m)} = D\mathbf{f}^{(m)}(\mathbf{z}^{(m)*})\Delta\mathbf{z}^{(m)} \qquad (2.54)$$

and the eigenvalue analysis of equilibrium $\mathbf{z}^*$ *is given by*

$$\left| D\mathbf{f}^{(m)}(\mathbf{z}^{(m)*}) - \lambda\mathbf{I}_{n(2N+1)\times n(2N+1)} \right| = 0, \qquad (2.55)$$

where $D\mathbf{f}^{(m)}(\mathbf{z}^{(m)*}) = \partial\mathbf{f}^{(m)}(\mathbf{z}^{(m)})/\partial\mathbf{z}^{(m)}\big|_{\mathbf{z}^{(m)*}}$. *The stability and bifurcation of periodic solution can be classified by the eigenvalues of* $D\mathbf{f}^{(m)}(\mathbf{z}^{(m)*})$ *with*

$$(n_1, n_2, n_3 | n_4, n_5, n_6). \qquad (2.56)$$

(i) *If all eigenvalues of the equilibrium possess negative real parts, the approximate periodic solution is stable.*

(ii) *If at least one of eigenvalues of the equilibrium possesses positive real part, the approximate periodic solution is unstable.*

(iii) *The boundaries between stable and unstable equilibriums with higher-order singularity give bifurcation and stability conditions with higher-order singularity.*

Proof. See Luo [12, 35, 36].   □

If $m \to \infty$, Eq. (2.45) will give the analytical expression of chaos in periodically excited, nonlinear dynamical systems in Eq. (2.32), which can be approximated by Eq. (2.48) under the condition of $||\mathbf{x}^{(m)}(t) - \mathbf{x}^{(m)*}(t)|| < \varepsilon$ as $N \to \infty$.

CHAPTER 3

Analytical Periodic Evolutions

In this chapter, the analytical solutions of periodic evolutions in the periodically diffused Brusselator will be presented.

3.1 CHEMICAL REACTION MODELS

As in Prigogine and Lefever [1], the autonomous Brusselator was proposed to characterize a set of four chemical reactions as follows:

$$
\begin{aligned}
A &\xrightarrow{k_1} X, \\
B + X &\xrightarrow{k_2} Y + D, \\
2X + Y &\xrightarrow{k_3} 3X, \\
X &\xrightarrow{k_4} E,
\end{aligned}
\tag{3.1}
$$

where A and B are reactants, X and Y are catalysts, and D and E are products. Constants $k_i (i = 1, 2, 3, 4)$ are reaction rates of each sub-step, respectively. $[A], [B], [D], [E], [X]$, and $[Y]$ are the concentrations of chemicals A, B, D, E, X, and Y, respectively. The capital letter "T" is time. The evolutions of concentrations in each sub-step are given in Table 3.1.

The global rate is given by summarization of rates in the sub-steps

$$
\begin{aligned}
\frac{d[A]}{dT} &= -k_1[A], \quad \frac{d[B]}{dT} = -k_2[B][X], \\
\frac{d[D]}{dT} &= k_2[B][X], \quad \frac{d[E]}{dT} = k_4[X], \\
\frac{d[X]}{dT} &= k_1[A] - k_2[B][X] + k_3[X]^2[Y] - k_4[X], \\
\frac{d[Y]}{dT} &= k_2[B][X] - k_3[X]^2[Y].
\end{aligned}
\tag{3.2}
$$

Table 3.1: Rate of concentration in sub-steps

Sub-Steps	Concentration Rates of Reactants	Concentration Rates of Products
$A \xrightarrow{k_1} X$	$\dfrac{d[A]}{dT} = -k_1[A]$	$\dfrac{d[X]}{dT} = k_1[A]$
$B + X \xrightarrow{k_2} Y + D$	$\dfrac{d[B]}{dT} = -k_2[B]\,[X]$	$\dfrac{d[Y]}{dT} = k_2[B]\,[X]$
	$\dfrac{d[X]}{dT} = -k_2[B]\,[X]$	$\dfrac{d[D]}{dT} = k_2[B]\,[X]$
$2X + Y \xrightarrow{k_3} 3X$	$\dfrac{d[X]}{dT} = -2k_3[X]^2[Y]$	$\dfrac{d[X]}{dT} = 3k_3[X]^2\,[Y]$
	$\dfrac{d[Y]}{dT} = -k_3[X]^2[Y]$	
$X \xrightarrow{k_4} E$	$\dfrac{d[X]}{dT} = -k_4[X]$	$\dfrac{d[E]}{dT} = k_4[X]$

Rescaling the concentrations and time gives

$$x = \sqrt{\frac{k_3}{k_4}}[X], \quad y = \sqrt{\frac{k_3}{k_4}}[Y], \quad t = k_4 T,$$

$$a = \frac{k_1}{k_4}\sqrt{\frac{k_3}{k_4}}[A], \quad b = \frac{k_2}{k_4}[B], \quad d = \sqrt{\frac{k_3}{k_4}}[D], \quad e = \sqrt{\frac{k_3}{k_4}}[E]. \tag{3.3}$$

With the forgoing variables, the simplified equation of Eq. (3.2) is

$$\frac{\partial a}{\partial t} = -\frac{k_1}{k_4}a, \quad \frac{\partial b}{\partial t} = -\frac{k_2}{\sqrt{k_3 k_4}}x, \quad \frac{\partial d}{\partial t} = bx,$$

$$\frac{\partial e}{\partial t} = x, \quad \frac{\partial x}{\partial t} = a - (b+1)x + x^2 y, \quad \frac{\partial y}{\partial t} = bx - x^2 y. \tag{3.4}$$

The foregoing system is an autonomous system with respect to the evolution of a, b, x, and y. The last two terms of Eq. (3.4) are sufficient to present the entire system. Thus, the governing differential equations for the autonomous Brusselator are

$$\dot{x} = a - (b+1)x + x^2 y,$$
$$\dot{y} = bx - x^2 y, \tag{3.5}$$

where $\dot{x} = dx/dt$ and $\dot{y} = dy/dt$. As in Tomita et al. [4], a harmonic diffusion term is introduced, and the Brusselator with a harmonic diffusion is

$$\dot{x} = a + (b + 1)x - x^2 y - Q_0 \cos \Omega t,$$
$$\dot{y} = bx + x^2 y,$$
$$(3.6)$$

where a and b are constant because constant input concentrations are usually wanted. Q_0 and Ω are excitation amplitude and frequency, respectively.

3.2   ANALYTICAL SOLUTIONS

Periodic evolutions of the system in Eq. (3.5) can studied by the generalized harmonic balance method. In Luo [12, 35, 36], the generalized format of a two-dimensional system is

$$\dot{\mathbf{x}} = \mathbf{f}(\mathbf{x}, t),$$
$$(3.7)$$

where

$$\mathbf{x} = (x, y)^{\mathrm{T}}, \quad \dot{\mathbf{x}} = (\dot{x}, \dot{y})^{\mathrm{T}}, \quad \mathbf{f} = (f_1, f_2)^{\mathrm{T}}$$
$$f_1 = a - (b + 1)x + x^2 y + Q_0 \cos \Omega t, \quad f_2 = bx - x^2 y.$$
$$(3.8)$$

The analytical solution of period-m evolution ($m = 1, 2, 3, \ldots$) for Eq. (3.7) is represented approximately by a finite Fourier series

$$x^{(m)*}(t) \approx a_{(1)0/m} + \sum_{l=1}^{N} b_{(1)l/m} \cos\left(\frac{l}{m}\Omega t\right) + c_{(1)l/m} \sin\left(\frac{l}{m}\Omega t\right),$$
$$y^{(m)*}(t) \approx a_{(1)0/m} + \sum_{l=1}^{N} b_{(2)l/m} \cos\left(\frac{l}{m}\Omega t\right) + c_{(2)l/m} \sin\left(\frac{l}{m}\Omega t\right).$$
$$(3.9)$$

The first-order derivatives of $x^{(m)*}(t)$ and $y^{(m)*}(t)$ are

$$\dot{x}^{(m)*}(t) \approx \dot{a}_{(1)0/m} + \sum_{l=1}^{N} \left(\dot{b}_{(1)l/m} + \frac{l\Omega c_{(1)l/m}}{m}\right) \cos\left(\frac{l}{m}\Omega t\right)$$
$$+ \left(\dot{c}_{(1)l/m} - \frac{l\Omega b_{(1)l/m}}{m}\right) \sin\left(\frac{l}{m}\Omega t\right),$$
$$\dot{y}^{(m)*}(t) \approx \dot{a}_{(2)0/m} + \sum_{l=1}^{N} \left(\dot{b}_{(2)l/m} + \frac{l\Omega c_{(2)l/m}}{m}\right) \cos\left(\frac{l}{m}\Omega t\right)$$
$$+ \left(\dot{c}_{(2)l/m} - \frac{l\Omega b_{(2)l/m}}{m}\right) \sin\left(\frac{l}{m}\Omega t\right).$$
$$(3.10)$$

Substitution of Eqs. (3.8) and (3.9) into Eq. (3.5) and averaging for both $\cos(l\Omega t/m)$ and $\sin(l\Omega t/m)$ terms ($l = 1, 2, \ldots N$) over the period mT ($T = 2\pi/\Omega$) gives

$$
\begin{aligned}
\dot{a}_{(1)0/m} &= F_{(1)0/m}(\mathbf{z}^{(m)}), \\
\dot{b}_{(1)l/m} &= -\frac{l}{m}\Omega c_{(1)l/m} + F^{(c)}_{(1)l/m}(\mathbf{z}^{(m)}), \\
\dot{c}_{(1)l/m} &= \frac{l}{m}\Omega b_{(1)l/m} + F^{(s)}_{(1)l/m}(\mathbf{z}^{(m)}), \\
\dot{a}_{(2)0/m} &= F_{(2)0/m}(\mathbf{z}^{(m)}), \\
\dot{b}_{(2)l/m} &= -\frac{l}{m}\Omega c_{(2)l/m} + F^{(c)}_{(2)l/m}(\mathbf{z}^{(m)}), \\
\dot{c}_{(2)l/m} &= \frac{l}{m}\Omega b_{(2)l/m} + F^{(s)}_{(2)l/m}(\mathbf{z}^{(m)}),
\end{aligned}
\tag{3.11}
$$

where

$$
\begin{aligned}
F_{(1)0/m}(\mathbf{z}^{(m)}) &= \frac{1}{mT}\int_0^{mT} f_1(x^{(m)*}, y^{(m)*}, t)\,dt, \\
F^{(c)}_{(1)l/m}(\mathbf{z}^{(m)}) &= \frac{2}{mT}\int_0^{mT} f_1(x^{(m)*}, y^{(m)*}, t)\cos\left(\frac{l}{m}\Omega t\right)dt, \\
F^{(s)}_{(1)l/m}(\mathbf{z}^{(m)}) &= \frac{2}{mT}\int_0^{mT} f_1(x^{(m)*}, y^{(m)*}, t)\sin\left(\frac{l}{m}\Omega t\right)dt, \\
F_{(2)0/m}(\mathbf{z}^{(m)}) &= \frac{1}{mT}\int_0^{mT} f_2(x^{(m)*}, y^{(m)*}, t)\,dt, \\
F^{(c)}_{(2)l/m}(\mathbf{z}^{(m)}) &= \frac{2}{mT}\int_0^{mT} f_2(x^{(m)*}, y^{(m)*}, t)\cos\left(\frac{l}{m}\Omega t\right)dt, \\
F^{(s)}_{(2)l/m}(\mathbf{z}^{(m)}) &= \frac{2}{mT}\int_0^{mT} f_2(x^{(m)*}, y^{(m)*}, t)\sin\left(\frac{l}{m}\Omega t\right)dt,
\end{aligned}
\tag{3.12}
$$

and the harmonic amplitude vector $\mathbf{z}^{(m)}$ is defined as

$$
\begin{aligned}
\mathbf{z}^{(m)} &\equiv \left(a_{(1)0/m}, \mathbf{b}_{(1)m}, \mathbf{c}_{(1)m}, a_{(2)0/m}, \mathbf{b}_{(2)m}, \mathbf{c}_{(2)m}\right)^{\mathrm{T}} \\
&= (z_1, z_2, \ldots, z_{2N+1}, z_{2N+2}, \ldots, z_{4N+2})^{\mathrm{T}}
\end{aligned}
\tag{3.13}
$$

with

$$
\begin{aligned}
\mathbf{b}_{(1)m} &\equiv \left(b_{(1)1/m}, \ldots, b_{(1)N/m}\right)^{\mathrm{T}}, \quad \mathbf{c}_{(1)m} \equiv \left(c_{(1)1/m}, \ldots, c_{(1)N/m}\right)^{\mathrm{T}} \\
\mathbf{b}_{(2)m} &\equiv \left(b_{(2)1/m}, \ldots, b_{(2)N/m}\right)^{\mathrm{T}}, \quad \mathbf{c}_{(2)m} \equiv \left(c_{(2)1/m}, \ldots, c_{(2)N/m}\right)^{\mathrm{T}}.
\end{aligned}
\tag{3.14}
$$

The detailed expression in Eq. (3.11) is given as follows. The constant term for concentration x is

$$F_{(1)0}^{(m)} = a - (b+1)a_{(1)0/m} + f_{01}^{(m)} + \frac{1}{2}\sum_{k=1}^{N} f_{01}^{(m)}(k) + \frac{1}{4}\sum_{i=1}^{N}\sum_{j=1}^{N}\sum_{k=1}^{N} f_{02}^{(m)}(i,j,k), \quad (3.15)$$

where

$$\begin{aligned}
f_{01}^{(m)} &= \left(a_{(1)0/m}\right)^2 a_{(2)0/m}, \\
f_{02}^{(m)}(k) &= a_{(2)0/m}\left(b_{(1)k/m}b_{(1)k/m} + c_{(1)k/m}c_{(1)k/m}\right) \\
&\quad + 2a_{(1)0/m}\left(b_{(1)k/m}b_{(2)k/m} + c_{(1)k/m}c_{(2)k/m}\right), \\
f_{03}^{(m)}(i,j,k) &= b_{(1)i/m}b_{(1)j/m}b_{(2)k/m}\Delta_1 \\
&\quad + \left(2b_{(1)i/m}c_{(1)j/m}c_{(1)k/m} + b_{(2)i/m}c_{(1)j/m}c_{(1)k/m}\right)\Delta_2,
\end{aligned} \quad (3.16)$$

and

$$\Delta_1 = \delta_{i+j-k}^0 + \delta_{i-j+k}^0 + \delta_{i-j-k}^0, \quad \Delta_2 = \delta_{i+j-k}^0 + \delta_{i-j+k}^0 - \delta_{i-j-k}^0. \quad (3.17)$$

The cosine term for concentration x is

$$\begin{aligned}
F_{(1)l/m}^{(c)} &= -(b+1)b_{(1)l/m} + Q_0\delta_m^l + f_{l/m1}^{(c)} \\
&\quad + \frac{1}{2}\sum_{i=1}^{N}\sum_{j=1}^{N} f_{l/m1}^{(c)}(i,j) + \frac{1}{4}\sum_{i=1}^{N}\sum_{j=1}^{N}\sum_{k=1}^{N} f_{l/m2}^{(c)}(i,j,k),
\end{aligned} \quad (3.18)$$

where

$$\begin{aligned}
f_{l/m1}^{(c)} &= \left(a_{(1)0/m}\right)^2 b_{(2)l/m}, \\
f_{l/m2}^{(c)}(i,j) &= \left(a_{(2)0/m}b_{(1)i/m}b_{(1)j/m} + 2a_{(1)0/m}b_{(1)i/m}b_{(2)j/m}\right)\Delta_{11} \\
&\quad + \left(a_{(2)0/m}c_{(1)i/m}c_{(1)j/m} + 2a_{(1)0/m}c_{(1)i/m}c_{(2)j/m}\right)\Delta_{12} \\
f_{l/m3}^{(c)}(i,j,k) &= b_{(1)i/m}b_{(1)j/m}b_{(2)k/m}\Delta_{13} \\
&\quad + \left(2b_{(1)i/m}c_{(1)j/m}c_{(2)k/m} + b_{(2)i/m}c_{(1)j/m}c_{(1)k/m}\right)\Delta_{14}
\end{aligned} \quad (3.19)$$

and

$$\begin{aligned}
\Delta_{11} &= \delta_{i+j}^l + \delta_{|i-j|}^l, \quad \Delta_{12} = -\delta_{i+j}^l + \delta_{|i-j|}^l, \\
\Delta_{13} &= \delta_{i+j+k}^l + \delta_{|i+j-k|}^l + \delta_{|i-j+k|}^l + \delta_{|i-j-k|}^l, \\
\Delta_{14} &= \delta_{|i+j-k|}^l - \delta_{i+j+k}^l + \delta_{|i-j+k|}^l - \delta_{|i-j-k|}^l.
\end{aligned} \quad (3.20)$$

The sine term for the concentration x is

$$F_{(1)l/m}^{(s)} = -(b+1)c_{(1)l/m} + f_{l/m1}^{(s)}$$
$$+ \frac{1}{2}\sum_{i=1}^{N}\sum_{j=1}^{N} f_{l/m2}^{(s)}(i,j) + \frac{1}{4}\sum_{i=1}^{N}\sum_{j=1}^{N}\sum_{k=1}^{N} f_{l/m3}^{(s)}(i,j,k), \tag{3.21}$$

where

$$f_{l/m1}^{(s)} = 2a_{(1)0/m}a_{(2)0/m}c_{(1)l/m} + \left(a_{(1)0/m}\right)^2 c_{(2)l/m},$$
$$f_{l/m2}^{(s)}(i,j) = \left(a_{(2)0/m}b_{(1)i/m}c_{(1)j/m} + a_{(1)0/m}b_{(2)i/m}c_{(1)j/m}\right.$$
$$\left. + a_{(1)0/m}b_{(1)i/m}c_{(2)j/m}\right)\Delta_{21}, \tag{3.22}$$
$$f_{l/m3}^{(s)}(i,j,k) = \left(2b_{(1)i/m}b_{(2)j/m}c_{(1)k/m} + b_{(1)i/m}b_{(1)j/m}c_{(2)k/m}\right)\Delta_{22}$$
$$+ c_{(1)i/m}c_{(1)j/m}c_{(2)k/m}\Delta_{23}$$

and

$$\Delta_{21} = \delta_{i+j}^{l} - \mathrm{sgn}(i-j)\delta_{|i-j|}^{l},$$
$$\Delta_{22} = \delta_{i+j+k}^{l} - \mathrm{sgn}(i+j-k)\delta_{|i+j-k|}^{l}$$
$$+ \mathrm{sgn}(i-j+k)\delta_{|i-j+k|}^{l} - \mathrm{sgn}(i-j-k)\delta_{|i-j-k|}^{l}, \tag{3.23}$$
$$\Delta_{23} = -\delta_{i+j+k}^{l} + \mathrm{sgn}(i+j-k)\delta_{|i+j-k|}^{l}$$
$$+ \mathrm{sgn}(i-j+k)\delta_{|i-j+k|}^{l} - \mathrm{sgn}(i-j-k)\delta_{|i-j-k|}^{l}.$$

The constant term for the concentration y is

$$F_{(2)0}^{(m)} = ba_{(1)/m} - f_{01}^{(m)} - \frac{1}{2}\sum_{k=1}^{N} f_{01}^{(m)}(k) - \frac{1}{4}\sum_{i=1}^{N}\sum_{j=1}^{N}\sum_{k=1}^{N} f_{02}^{(m)}(i,j,k). \tag{3.24}$$

The cosine term for the concentration y is

$$F_{(2)l/m}^{(c)} = bb_{(1)l/m} - f_{l/m1}^{(c)} - \frac{1}{2}\sum_{i=1}^{N}\sum_{j=1}^{N} f_{l/m1}^{(c)}(i,j) - \frac{1}{4}\sum_{i=1}^{N}\sum_{j=1}^{N}\sum_{k=1}^{N} f_{l/m2}^{(c)}(i,j,k). \tag{3.25}$$

The sine term for the concentration y is

$$F_{(2)l/m}^{(s)} = bc_{(1)l/m} - f_{l/m1}^{(s)} - \frac{1}{2}\sum_{i=1}^{N}\sum_{j=1}^{N} f_{l/m2}^{(s)}(i,j) - \frac{1}{4}\sum_{i=1}^{N}\sum_{j=1}^{N}\sum_{k=1}^{N} f_{l/m3}^{(s)}(i,j,k). \tag{3.26}$$

Using the definition of Eq. (3.12), Eq. (3.10) becomes

$$\dot{\mathbf{z}}^{(m)} = \mathbf{g}^{(m)}(\mathbf{z}^{(m)}), \tag{3.27}$$

where

$$g^{(m)}(\mathbf{z}^{(m)}) = \Big(F_{(1)0}(\mathbf{z}^{(m)}), \quad -\frac{\Omega}{m}\mathbf{k}\mathbf{c}_{(1)m} + \mathbf{F}_{(1)c}(\mathbf{z}^{(m)}), \quad \frac{\Omega}{m}\mathbf{k}\mathbf{b}_{(1)m} + \mathbf{F}_{(1)s}(\mathbf{z}^{(m)}),$$
$$F_{(2)0}(\mathbf{z}^{(m)}), \quad -\frac{\Omega}{m}\mathbf{k}\mathbf{c}_{(2)m} + \mathbf{F}_{(2)c}(\mathbf{z}^{(m)}), \quad \frac{\Omega}{m}\mathbf{k}\mathbf{b}_{(2)m} + \mathbf{F}_{(2)s}(\mathbf{z}^{(m)}) \Big)^{\mathrm{T}} \tag{3.28}$$

and

$$k = diag(1, 2, \ldots, N),$$
$$\mathbf{F}_{(1)c}(\mathbf{z}^{(m)}) = \Big(F^{(c)}_{(1)1/m}(\mathbf{z}^{(m)}), \ldots, F^{(c)}_{(1)N/m}(\mathbf{z}^{(m)}) \Big)^{\mathrm{T}},$$
$$\mathbf{F}_{(1)s}(\mathbf{z}^{(m)}) = \Big(F^{(s)}_{(1)1/m}(\mathbf{z}^{(m)}), \ldots, F^{(s)}_{(1)N/m}(\mathbf{z}^{(m)}) \Big)^{\mathrm{T}},$$
$$\mathbf{F}_{(2)c}(\mathbf{z}^{(m)}) = \Big(F^{(c)}_{(2)1/m}(\mathbf{z}^{(m)}), \ldots, F^{(c)}_{(2)N/m}(\mathbf{z}^{(m)}) \Big)^{\mathrm{T}},$$
$$\mathbf{F}_{(2)s}(\mathbf{z}^{(m)}) = \Big(F^{(s)}_{(2)1/m}(\mathbf{z}^{(m)}), \ldots, F^{(s)}_{(2)N/m}(\mathbf{z}^{(m)}) \Big)^{\mathrm{T}}. \tag{3.29}$$

The period-m evolution of the periodically force Bruseelator can be obtained by setting $\dot{\mathbf{z}}^{(m)} = \mathbf{0}$, i.e.,

$$F_{(1)0/m}(\mathbf{z}^{(m)*}) = 0, \quad -\frac{\Omega}{m}\mathbf{k}\mathbf{c}^*_{(1)m} + \mathbf{F}_{(1)c}(\mathbf{z}^{(m)*}) = \mathbf{0}, \quad \frac{\Omega}{m}\mathbf{k}\mathbf{c}^*_{(1)m} + \mathbf{F}_{(1)s}(\mathbf{z}^{(m)*}) = \mathbf{0},$$
$$F_{(2)0/m}(\mathbf{z}^{(m)*}) = 0, \quad -\frac{\Omega}{m}\mathbf{k}\mathbf{c}^*_{(2)m} + \mathbf{F}_{(2)c}(\mathbf{z}^{(m)*}) = \mathbf{0}, \quad \frac{\Omega}{m}\mathbf{k}\mathbf{b}^*_{(2)m} + \mathbf{F}_{(2)s}(\mathbf{z}^{(m)}) = \mathbf{0}. \tag{3.30}$$

The nonlinear equations of Eq. (3.29) can be solved by the Newton-Raphson method. In Luo [12, 21, 22], the linearized equation in the vicinity of equilibrium $\mathbf{z}^{(m)*}$ is given by

$$\Delta\dot{\mathbf{z}}^{(m)} = D\mathbf{g}^{(m)}(\mathbf{z}^{(m)*})\Delta\mathbf{z}^{(m)}, \tag{3.31}$$

where

$$D\mathbf{g}^{(m)}(\mathbf{z}^{(m)*}) = \frac{\partial \mathbf{g}^{(m)}(\mathbf{z}^{(m)})}{\partial \mathbf{z}^{(m)}}\bigg|_{\mathbf{z}^{(m)}=\mathbf{z}^{(m)*}} = \left(\frac{\partial g_i}{\partial z_j}\right)_{2(2N+1)\times 2(2N+1)}, \tag{3.32}$$

where for $r = 1, 2, \ldots, 4N + 2$ and $l = 1, 2, \ldots, N$, we have

$$\frac{\partial g_1}{\partial z_r} = \frac{\partial F_{(1)0}(\mathbf{z}^{(m)})}{\partial z_r} = g^{(0)}_{(1)r},$$

$$\frac{\partial g_{l+1}}{\partial z_r} = \frac{\partial F^{(c)}_{(1)l/m}(\mathbf{z}^{(m)})}{\partial z_r} = g^{(c)}_{(1)lr},$$

$$\frac{\partial g_{N+1+l}}{\partial z_r} = \frac{\partial F^{(s)}_{(1)l/m}(\mathbf{z}^{(m)})}{\partial z_r} = g^{(s)}_{(1)lr},$$

$$\frac{\partial g_{2N+2}}{\partial z_r} = \frac{\partial F_{(2)0}(\mathbf{z}^{(m)})}{\partial z_r} = g^{(0)}_{(2)r}, \qquad (3.33)$$

$$\frac{\partial g_{2N+2+l}}{\partial z_r} = \frac{\partial F^{(c)}_{(2)l/m}(\mathbf{z}^{(m)})}{\partial z_r} = g^{(c)}_{(2)lr},$$

$$\frac{\partial g_{3N+2+l}}{\partial z_r} = \frac{\partial F^{(s)}_{(2)l/m}(\mathbf{z}^{(m)})}{\partial z_r} = g^{(s)}_{(1)lr}.$$

Thus, the derivative of the constant term of the concentration x for $r = 1, 2, \ldots, 4N + 2$ is

$$g^{(0)}_{(1)r} = (b+1)\delta^{r-1}_0 + g^{(0)}_{r1} + \frac{1}{2}\sum_{k=1}^{N} g^{(0)}_{r2}(k) + \frac{1}{4}\sum_{i=1}^{N}\sum_{j=1}^{N}\sum_{k=1}^{N} g^{(0)}_{r3}(i,j,k), \qquad (3.34)$$

where

$$g^{(0)}_{r1} = 2\delta^{r-1}_0 a_{(1)0/m} a_{(2)0/m} + \delta^{r-1}_{2N+1}\left(a_{(1)0/m}\right)^2 a_{(2)0/m}, \qquad (3.35)$$

$$\begin{aligned}
g^{(0)}_{r2}(k) &= \delta^{r-1}_{2N+1}\left(b_{(1)k/m}b_{(1)k/m} + c_{(1)k/m}c_{(1)k/m}\right) \\
&\quad + 2a_{(2)0/m}\left(\delta^{r-1}_k b_{(1)k/m} + \delta^{r-1}_{k+N}c_{(1)k/m}\right) \\
&\quad + 2\delta^{r-1}_0\left(b_{(1)k/m}b_{(2)k/m} + c_{(1)k/m}c_{(2)k/m}\right) \\
&\quad + 2a^{(m)}_{(1)0}\left[\delta^{r-1}_k b_{(2)k/m} + \delta^{r-1}_{k+2N+1}b_{(1)k/m}\right. \\
&\quad \left. + \delta^{r-1}_{k+N}c_{(2)k/m} + \delta^{r-1}_{k+3N+1}c_{(1)k/m}\right],
\end{aligned} \qquad (3.36)$$

$$\begin{aligned}
g^{(0)}_{r3}(i,j,k) &= \left[\delta^{r-1}_i b_{(1)j/m}b_{(2)k/m} + \delta^{r-1}_j b_{(1)i/m}b_{(2)k/m} + \delta^{r-1}_{k+2N+1}b_{(1)i/m}b_{(1)j/m}\right]\Delta_1 \\
&\quad + \left[2\delta^{r-1}_i c_{(1)j/m}c_{(1)k/m} + \delta^{r-1}_{j+N}b_{(1)i/m}c_{(1)k/m} + \delta^{r-1}_{k+N}b_{(1)i/m}c_{(1)j/m}\right. \\
&\quad \left. + \delta^{r-1}_{i+2N+1}c_{(1)j/m}c_{(1)k/m} + \delta^{r-1}_{j+N}b_{(2)i/m}c_{(1)k/m} + \delta^{r-1}_{k+N}b_{(2)i/m}c_{(1)j/m}\right]\Delta_2.
\end{aligned} \qquad (3.37)$$

The derivative of the cosine term for the concentration x for $r = 1, 2, \ldots, 4N + 2$ is

$$g_{(1)lr}^{(c)} = -\frac{l\Omega}{m}\delta_{l+N}^{r-1} - (b+1)\delta_l^{r-1} + g_{lr1}^{(c)} + \frac{1}{2}\sum_{i=1}^{N}\sum_{j=1}^{N}g_{lr2}^{(c)}(i,j) + \frac{1}{4}\sum_{i=1}^{N}\sum_{j=1}^{N}\sum_{k=1}^{N}g_{lr3}^{(c)}(i,j,k),$$

(3.38)

where

$$g_{lr1}^{(c)} = 2\delta_0^{r-1}a_{(1)0/m}b_{(2)l/m} + \delta_{l+2N+1}^{r-1}\left(a_{(1)0/m}\right)^2,$$

(3.39)

$$
\begin{aligned}
g_{lr2}^{(c)}(i,j) = &\left[\left(\delta_{2N+1}^{r-1}b_{(1)i/m}b_{(1)j/m} + \delta_i^{r-1}a_{(1)0/m}b_{(1)j/m} + \delta_j^{r-1}a_{(2)0/m}b_{(1)i/m}\right)\right.\\
&\left.+2\left(\delta_0^{r-1}b_{(1)i/m}b_{(2)j/m} + \delta_i^{r-1}a_{(1)0/m}b_{(2)j/m} + \delta_{j+2N+1}^{r-1}a_{(1)0/m}b_{(1)i/m}\right)\right]\Delta_{11}\\
&+\left[\left(2\delta_{2N+1}^{r-1}c_{(1)i/m}c_{(1)j/m} + \delta_{i+N}^{r-1}a_{(2)0/m}c_{(1)j/m} + \delta_{j+N}^{r-1}a_{(2)0/m}c_{(1)j/m}\right)\right.\\
&\left.+2\left(\delta_0^{r-1}c_{(1)i/m}c_{(2)j/m} + \delta_{i+N}^{r-1}a_{(1)0/m}c_{(2)j/m} + \delta_{j+3N+1}^{r-1}a_{(1)0/m}c_{(1)i/m}\right)\right]\Delta_{12},
\end{aligned}
$$

(3.40)

$$
\begin{aligned}
g_{lr3}^{(c)}(i,j,k) = &\left(\delta_i^{r-1}b_{(1)j/m}b_{(2)k/m} + \delta_j^{r-1}b_{(1)i/m}b_{(2)k/m} + \delta_{k+2N+1}^{r-1}b_{(1)i/m}b_{(1)j/m}\right)\Delta_{13}\\
&+\left[2\left(\delta_i^{r-1}c_{(1)j/m}c_{(2)k/m} + \delta_{j+N}^{r-1}b_{(1)i/m}c_{(2)k/m} + \delta_{k+3N+1}^{r-1}b_{(1)i/m}c_{(1)j/m}\right)\right.\\
&\left.+\left(\delta_{i+2N+1}^{r-1}c_{(1)j/m}c_{(1)k/m} + \delta_{j+N}^{r-1}b_{(2)i/m}c_{(1)k/m} + \delta_{k+N}^{r-1}b_{(2)i/m}c_{(1)j/m}\right)\right]\Delta_{14}.
\end{aligned}
$$

(3.41)

The derivative of the sine term of the concentration x for $r = 1, 2, \ldots, 4N + 2$ is

$$g_{(1)lr}^{(s)} = \frac{l\Omega}{m}\delta_l^{r-1} - (b+1)\delta_{l+3N+1}^{r-1} + g_{lr1}^{(s)} + \frac{1}{2}\sum_{i=1}^{N}\sum_{j=1}^{N}g_{lr2}^{(s)}(i,j)$$

$$+ \frac{1}{4}\sum_{i=1}^{N}\sum_{j=1}^{N}\sum_{k=1}^{N}g_{lr3}^{(s)}(i,j,k),$$

(3.42)

where

$$
\begin{aligned}
g_{lr1}^{(s)} = &2\left(\delta_0^{r-1}a_{(2)0/m}c_{(1)l/m} + \delta_{2N+1}^{r-1}a_{(1)0/m}c_{(1)l/m} + \delta_{l+N}^{r-1}a_{(1)0/m}a_{(2)0/m}\right)\\
&+ 2\delta_0^{r-1}a_{(1)0/m}c_{(2)l/m} + \delta_{l+3N+1}^{r-1}\left(a_{(1)0/m}\right)^2,
\end{aligned}
$$

(3.43)

$$
\begin{aligned}
g_{lr2}^{(s)}(i,j) = &\left[\left(\delta_{2N+1}^{r-1}b_{(1)i/m}c_{(1)j/m} + \delta_i^{r-1}a_{(2)0/m}c_{(1)j/m} + \delta_{j+N}^{r-1}a_{(2)0/m}b_{(1)i/m}\right)\right.\\
&+\left(\delta_0^{r-1}b_{(2)i/m}c_{(1)j/m} + \delta_{i+2N+1}^{r-1}a_{(1)0/m}c_{(1)j/m} + \delta_{j+N}^{r-1}a_{(1)0}^{(m)}b_{(2)i/m}\right)\\
&\left.+\left(\delta_0^{r-1}b_{(1)i/m}c_{(2)j/m} + \delta_i^{r-1}a_{(1)0/m}c_{(2)j/m} + \delta_{j+3N+1}^{r-1}a_{(1)0/m}b_{(1)i/m}\right)\right]\Delta_{21},
\end{aligned}
$$

(3.44)

$$
\begin{aligned}
g_{lr3}^{(s)}(i,j,k) = & \left[2\left(\delta_i^{r-1}b_{(2)j/m}c_{(1)k/m} + \delta_{j+2N+1}^{r-1}b_{(1)i/m}c_{(1)k/m} + \delta_{k+N}^{r-1}b_{(1)i/m}b_{(2)j/m}\right)\right. \\
& + \left(\delta_i^{r-1}b_{(1)j/m}c_{(2)k/m} + \delta_j^{r-1}b_{(1)i/m}c_{(2)k/m} + \delta_{k+3N+1}^{r-1}b_{(1)i/m}b_{(1)j/m}\right)\Big]\Delta_{22} \\
& + \left[\delta_{i+N}^{r-1}c_{(1)j/m}c_{(2)k/m} + \delta_{j+N}^{r-1}c_{(1)i/m}c_{(2)k/m} + \delta_{k+3N+1}^{r-1}c_{(1)i/m}c_{(1)j/m}\right]\Delta_{23}.
\end{aligned}
\tag{3.45}
$$

The derivative of the constant term of the concentration y is for $r = 1, 2, \ldots, 4N + 2$

$$
g_{(2)r}^{(0)} = b\delta_0^{r-1} - g_{r1}^{(0)} - \frac{1}{2}\sum_{k=1}^{\infty} g_{r2}^{(0)}(k) - \frac{1}{4}\sum_{i=1}^{\infty}\sum_{j=1}^{\infty}\sum_{k=1}^{\infty} g_{r3}^{(0)}(i,j,k).
\tag{3.46}
$$

The derivative of the cosine term of the concentration y for $r = 1, 2, \ldots, 4N + 2$ is

$$
\begin{aligned}
g_{(2)lr}^{(c)} = & -\frac{l\Omega}{m}\delta_{l+3N+1}^{r-1} + b\delta_l^{r-1} - g_{lr1}^{(c)} \\
& - \frac{1}{2}\sum_{i=1}^{\infty}\sum_{j=1}^{\infty} g_{lr2}^{(c)}(i,j) - \frac{1}{4}\sum_{i=1}^{\infty}\sum_{j=1}^{\infty}\sum_{k=1}^{\infty} f_{lr3}^{(c)}(i,j,k).
\end{aligned}
\tag{3.47}
$$

The derivative for the sine term of the concentration y for $r = 1, 2, \ldots, 4N + 2$ is

$$
\begin{aligned}
g_{(2)lr}^{(s)} = & \frac{l\Omega}{m}\delta_{l+2N+1}^{r-1} + b\delta_{l+N}^{r} - g_{lr1}^{(s)} \\
& - \frac{1}{2}\sum_{i=1}^{\infty}\sum_{j=1}^{\infty} g_{lr2}^{(s)}(i,j) - \frac{1}{4}\sum_{i=1}^{\infty}\sum_{j=1}^{\infty}\sum_{k=1}^{\infty} g_{lr3}^{(s)}(i,j,k).
\end{aligned}
\tag{3.48}
$$

The corresponding eigenvalues are determined by

$$
\left| D\mathbf{g}^{(m)}(\mathbf{z}^{(m)*}) - \lambda \mathbf{I}_{2(2N+1)\times 2(2N+1)} \right| = 0.
\tag{3.49}
$$

If $\mathrm{Re}\,\lambda_i < 0$ for $i = 1, 2, \ldots, 4N + 2$, the period-m solution is stable. If $\mathrm{Re}\,\lambda_i > 0$ for $i \in \{1, 2, \ldots, 4N + 2\}$, the period-$m$ solution is unstable. The boundary between stable and unstable solutions of the period-m evolution with high singularity gives bifurcation conditions of the periodic evolution.

CHAPTER 4

Analytical Routes to Chaotic Evolutions

As in Luo and Guo [38], the analytical solutions of periodic evolution of the Brusselator model should have been represented by an infinite Fourier expansion, which can avoid truncation error. However, such an infinite expansion cannot be applied in computation. Thus, the applicability of analytical solution necessitates the truncation on an infinite Fourier expansion, which is given by Eq. (3.10). The accuracy of the analytical solution thus depends on the truncation error, in other words depends on the number of harmonic terms preserved in Eq. (3.10).

4.1 BIFURCATION TREE BASED ON EXCITATION FREQUENCY

Once the number of harmonic terms prescribed in Eq. (3.10) is chosen to achieve a reasonable accuracy, all harmonic amplitudes varying with the diffusing frequency Ω can be computed by the method as presented in the previous section. To obtain harmonic amplitudes, Eq. (3.8) can be rewritten as

$$
\begin{aligned}
x^{(m)*}(t) &\approx a_{(1)0/m} + \sum_{l=1}^{N} A_{(1)l/m} \sin\left(\frac{l}{m}\Omega t + \varphi_{(1)l/m}\right), \\
y^{(m)*}(t) &\approx a_{(2)0/m} + \sum_{l=1}^{N} A_{(2)l/m} \sin\left(\frac{l}{m}\Omega t + \varphi_{(2)l/m}\right),
\end{aligned}
\tag{4.1}
$$

where

$$
\begin{aligned}
A_{(1)l/m} &= \sqrt{b_{(1)l/m}^2 + c_{(1)l/m}^2}, \quad \varphi_{(1)l/m} = \arctan\frac{b_{(1)l/m}}{c_{(1)l/m}}, \\
A_{(2)l/m} &= \sqrt{b_{(2)l/m}^2 + c_{(2)l/m}^2}, \quad \varphi_{(2)l/m} = \arctan\frac{b_{(2)l/m}}{c_{(2)l/m}}.
\end{aligned}
\tag{4.2}
$$

Without loss of generality, set the rate constant k_i to be unit ($i = 1, 2, 3, 4$). Thus, $a, b, x,$ and y are real concentration and t is real time. Consider a set of system parameters as

$$
a = 0.4, \quad b = 1.2, \quad Q_0 = 0.08.
\tag{4.3}
$$

Table 4.1: Bifurcation summary ($a = 0.4$, $b = 1.2$, $Q_0 = 0.08$)

Ω	Evolution Switching	Bifurcation
0.053283	unstable P-1 to stable P-1	SN
0.054307	stable P-1 to unstable P-1	SN
0.413493	P-1 to P-2	HB
0.736829	P-2 to P-4	HB
0.78681	P-4 to P-8	HB
0.7947	P-8 to P-16	HB
0.9199	P-16 to P-8	HB
0.92512	P-8 to P-4	HB
0.945221	P-4 to P-2	HB
1.023424	P-2 to P-1	HB
1.25	stable P-1 to unstable P-1	HB

The analytical approximate solutions of the Brusselator with a harmonic diffusion are presented for the harmonic amplitude vs. diffusion frequency. For all the following curves, the acronym "HB" is used to represent the Hopf bifurcation. The acronym "SN" is short for Saddle-node bifurcations. The "P-m," where m is an integer, indicates a period-m evolution. The stable periodic evolutions and unstable periodic evolutions are plotted in black solid lines and red dashed lines, respectively.

In Fig. 4.1, the harmonic amplitudes of evolutions of concentration x in the Brusselator in Eq. (3.6) are depicted in the diffusion frequency range $\Omega \in (0, 1.4]$. The constant term $a_{(1)0/m}$ vs. diffusion frequency Ω is not shown herein because it keeps constant. From Eq. (3.6), the constant term $a_{(1)0/m}$ must satisfy the following equations:

$$
\begin{aligned}
\dot{a}_{(1)0/m} &= a - (b + 1)a_{(1)0/m} + (a_{(1)0/m})^2 a_{(2)0/m} + h(\mathbf{z}^{(m)}), \\
\dot{a}_{(2)0/m} &= ba_{(1)0/m} - (a_{(1)0/m})^2 a_{(2)0/m} - h(\mathbf{z}^{(m)}).
\end{aligned}
\tag{4.4}
$$

Adding the two equations in Eq. (4.4) and applying periodicity $\dot{a}_{(1)0/m} = \dot{a}_{(2)0/m} = 0$, there exists

$$
a_{(1)0/m} = a. \tag{4.5}
$$

To discuss the bifurcation, the constant $a_{(2)0/m}$ varying with diffusion frequency is presented in Fig. 4.1. The constant $a_{(1)0/m} = a$ will not be presented herein. A summary of all bifurcations with diffusion frequency increase is tabulated in Table 4.1. As the harmonic diffusion starts with a small frequency very close to zero, the period-1 diffusion has a saddle-node

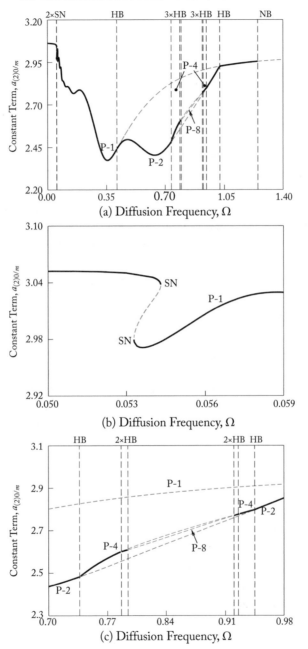

Figure 4.1: Constant $a_{(2)0/m}$ for concentration y varying with diffusion frequency: (a) global view ($\Omega \in (0, 1.4)$), (b) zoomed view ($\Omega \in (0.05, 0.06)$), and (c) zoomed view ($\Omega \in (0.70, 0.98)$). ($a = 0.4$, $b = 1.2$, $Q_0 = 0.08$).

bifurcation at $\Omega \approx 0.053283$, which is also for the onset of an unstable period-1 evolution. The unstable period-1 evolution disappears at $\Omega \approx 0.054307$ which is another saddle-node bifurcation. The zoomed views of the frequency-amplitude characteristics are presented in Fig. 4.1(b). A bifurcation tree of period-1 to period-8 motion is in $\Omega \in (0.413493, 1.023424)$, as shown in Fig. 4.1(c). The bifurcation tree of period-1 to period-8 motions is clearly presented through the constant evolutions, and the constants increase with diffusion frequency. The chaotic evolution of the Brusselator should exist in $\Omega \in (0.78681, 0.79470)$. The two Hopf bifurcations at $\Omega \approx 0.413493$ and 1.023424 are for the onsets of period-2 evolutions. The period-2 motion has two Hopf bifurcations of $\Omega \approx 0.736829$ and 0.945221, from which the period-4 evolution will appear. The period-4 evolution exists in $\Omega \in (0.736829, 0.945224)$. Such a period-4 evolution has two Hopf bifurcations of $\Omega \approx 0.78681$ and 0.92512 where the period-8 evolution appears. Similarly, a period-8 evolution exists in $\Omega \in (0.78681, 0.92512)$. The period-8 evolution has two Hopf bifurcations at $\Omega \approx 0.7947$ and 0.9199, which are the onset of period-16 evolutions. Herein, period-16 evolution will be presented because the stable period-16 evolution exists in the short frequency range. The period-1 to period-8 bifurcation tree provides an insight how the period-1 evolution to chaos. At $\Omega \approx 1.25$, another Hopf bifurcation of the period-1 evolution generates quasi-periodic evolutions.

To demonstrate the main harmonics effects on the periodic evolutions, the harmonic amplitudes $A_{(1)k}$ ($k = 1, 2, 3, 9, 10$) for period-m evolutions ($m = 1, 2, 4, 8$) are presented in Fig. 4.2. For the first-order primary harmonic amplitude $A_{(1)1}$, the global and zoomed views of the bifurcation trees are presented in Fig. 4.2(a)–(c). For the primary harmonic amplitudes of $A_{(1)2}$, $A_{(1)3}$, and $A_{(1)9}$, only the global view of the bifurcation tree are presented in Fig. 4.2(d)–(f), respectively. In Fig. 4.2(a), the quantity level of the first primary harmonic amplitude $A_{(1)1}$ is about $A_{(1)1} \sim 0.37$. In Fig. 4.2(b), $A_{(1)1}$ lies in $A_{(1)1} \in (0.08, 0.88)$ for $\Omega \in (0, 0.06)$. In Fig. 4.2(c), the main portion of bifurcation tree is in the frequency range of $\Omega \in (0.7, 1.05)$, and the corresponding harmonic amplitudes decay from $A_{(1)1} \approx 0.2$ to 0.08 for $\Omega \in (0.7, 1.05)$. The bifurcation tree of period-1 to period-8 motion is also observed clearly in Fig. 4.2(c). Thus, the harmonic amplitude of the chaotic diffusion should have the similar quantity level. To compare the quantity level changing with harmonic order, the second-order primary harmonic amplitude varying with diffusion frequency is presented in Fig. 4.2(d) for the bifurcation tree of period-1 to period-8 evolution. For $\Omega \in (0, 0.06)$, the second primary harmonic amplitude is in the range of $A_{(1)2} \in (10^{-4}, 5 \times 10^{-2})$. For $\Omega \in (0.35, 1.40)$, the amplitude quantity level of the second-order harmonic terms for the bifurcation tree is from $A_{(1)2} \approx 10^{-1}$ decaying to 10^{-3}. The third-order primary harmonic amplitude vs. diffusion frequency is presented in Fig. 4.2(e). The bifurcation tree based on the third-order primary harmonic amplitude is similar to the second-order primary harmonic amplitude. For $\Omega \in (0, 0.06)$, the third-order primary harmonic amplitude is in the range of $A_{(1)3} \in (10^{-5}, 10^{-2})$. For $\Omega \in (0.35, 1.40)$, the amplitude quantity level of the third harmonic terms for the bifurcation tree is from $A_{(1)3} \approx 10^{-1}$ decaying to 10^{-4}. To avoid abundant illustrations, the ninth-order primary harmonic amplitude vs. diffusion fre-

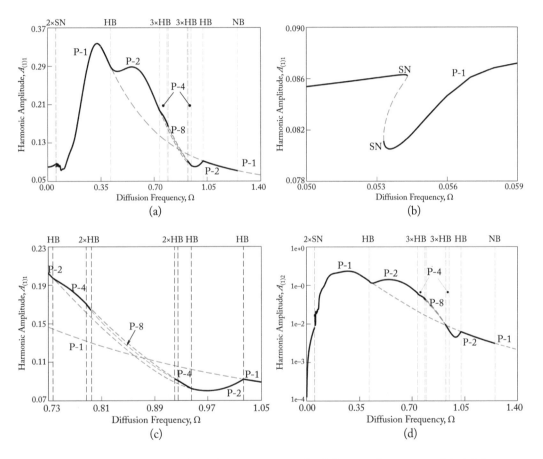

Figure 4.2: Harmonic amplitudes of x varying with diffusion frequency: (a)–(c) harmonic amplitude $A_{(1)1}$; (d)–(f) harmonic amplitudes $A_{(1)2}, A_{(1)3}, A_{(1)9}$. Parameter ($a = 0.4$, $b = 1.2$, $Q_0 = 0.08$). (*Continues.*)

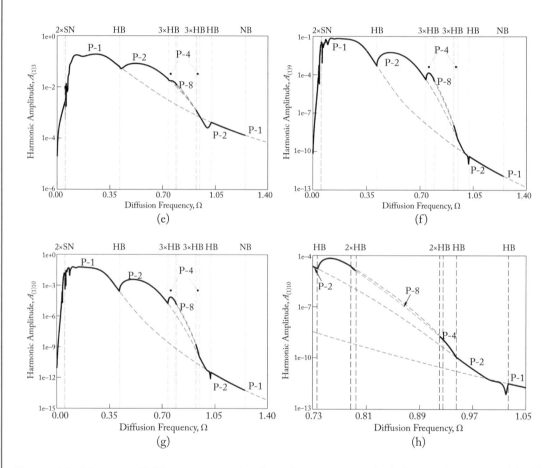

Figure 4.2: (*Continued.*) Harmonic amplitudes of x varying with diffusion frequency: (d)–(f) harmonic amplitudes $A_{(1)2}$, $A_{(1)3}$, $A_{(1)9}$; (g)–(h) harmonic amplitudes $A_{(1)10}$. Parameter ($a = 0.4$, $b = 1.2$, $Q_0 = 0.08$).

quency is placed in Fig. 4.2(f). For $\Omega \in (0, 0.06)$, the ninth primary harmonic amplitude is in $A_{(1)9} \in (10^{-10}, 10^{-2})$. For $\Omega \in (0.35, 1.40)$, the amplitude quantity level of the ninth harmonic terms for the bifurcation tree is from $A_{(1)9} \approx 10^{-1}$ decaying to 10^{-13}. With harmonic order increase, the decay rate of the harmonic amplitude increase with diffusion frequency. To end the illustration of harmonic amplitude, the tenth-order primary harmonic amplitude varying with diffusion amplitude is presented in Fig. 4.2(g)–(h). For the tenth-order primary harmonic amplitudes $A_{(1)10}$, the quantity level of the tenth primary harmonic amplitude $A_{(1)10}$ is with $A_{(1)10} \sim 10^{-2}$. In Fig. 4.2(h), $A_{(1)10} \in (10^{-12}, 5 \times 10^{-2})$ for $\Omega \in (0, 0.06)$. The main bifurcation tree amplitudes decay from $A_{(1)10} \approx 5 \times 10^{-2}$ to 10^{-12} for $\Omega \in (0.7, 1.05)$.

In the similar fashion, the harmonic amplitudes of the bifurcation tree of period-1 to period-8 evolution for concentration y is presented in Fig. 4.3(a)–(h). The bifurcation trees of the concentration y for different harmonic orders are similar to the concentration x. The quantity levels of the harmonic amplitudes for the different orders are lower than the concentration x. The detailed discussion will not be presented herein.

Besides the primary harmonic terms effects to the period-m evolution, the harmonic amplitudes of fractional orders are also very important. For the period-1 evolution, all fractional order harmonic terms are with zero amplitude. For the period-m evolutions, $\mathrm{mod}(k, m) \neq 0$. To avoid abundant illustration, only the harmonic amplitudes of three fractional orders $(A_{(1)1/8}, A_{(1)1/4}, A_{(1)1/2})$ for $k/m \in (0, 1)$ and three fractional orders $(A_{(1)19/2}, A_{(1)39/4}, A_{(1)79/8})$ for $k/m \in (9, 10)$ are presented in Fig. 4.4(a)–(f) for the concentration x. The harmonic amplitudes of $A_{(1)1/8}$ and $A_{(1)79/8}$ are for the period-8 evolution. For period-4, period-2, and period-1 evolutions, $A_{(1)1/8} = 0$ and $A_{(1)79/8} = 0$. So the bifurcation tree based on the 1/8 and 97/8-order harmonic amplitudes are very simple because the higher-order periodic evolutions until chaos were not presented, as shown in Fig. 4.4(a) and (f). For onsets of period-8 evolutions, the saddle-node bifurcations exist, which are the Hopf bifurcations for the period-4 evolution. For the period-8 evolution, $A_{(1)1/8} \sim 10^{-2}$ and $A_{(1)79/8} \sim 10^{-5}$. With diffusion frequency, $A_{(1)79/8}$ decays to $A_{(1)79/8} \sim 10^{-9}$. The harmonic amplitudes of $A_{(1)1/4}$ and $A_{(1)39/4}$ are presented in Fig. 4.4(b) and (e) for the period-8 and period-4 evolutions. For period-2 and period-1 evolutions, $A_{(1)1/4} = 0$ and $A_{(1)39/4} = 0$. For the period-8 and period-4 evolutions, $A_{(1)1/4} \sim 7.5 \times 10^{-2}$ and $A_{(1)39/4} \sim 10^{-4}$. With diffusion frequency, $A_{(1)39/4}$ decays to $A_{(1)39/4} \sim 10^{-10}$. The bifurcation trees based on the 1/4 and 39/4-order harmonic amplitudes are for the two period-8 and period-4 evolutions. The bifurcation trees possess two curves to represent the period-4 and period-8 evolutions. Again, the onsets of period-4 evolutions are the saddle-node bifurcations, which are also for the Hopf bifurcation of the period-2 evolutions. In Fig. 4.4(c) and (f), the harmonic amplitudes of $A_{(1)1/2}$ and $A_{(1)19/2}$ are for the period-8, period-4 and period-2 evolutions. The bifurcation tree has three curves for period-2, period-4 and period-8 evolutions. For the period-1 evolution, $A_{(1)1/2} = 0$ and $A_{(1)19/2} = 0$. For the period-8, period-4, and period-2 evolutions, $A_{(1)1/2} \sim 2.7 \times 10^{-1}$ and $A_{(1)19/2} \sim 10^{-3}$. With diffusion frequency, $A_{(1)19/2}$ decays to $A_{(1)19/2} \sim 10^{-11}$.

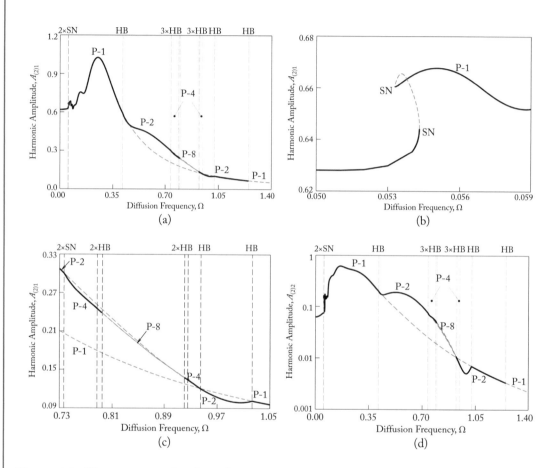

Figure 4.3: Harmonic amplitudes of y varying with diffusion frequency: (a)–(c) harmonic amplitude $A_{(2)1}$; (d)–(f) harmonic amplitude $A_{(2)2}$, $A_{(2)3}$, $A_{(2)9}$. Parameter ($a = 0.4$, $b = 1.2$, $Q_0 = 0.08$). (*Continues.*)

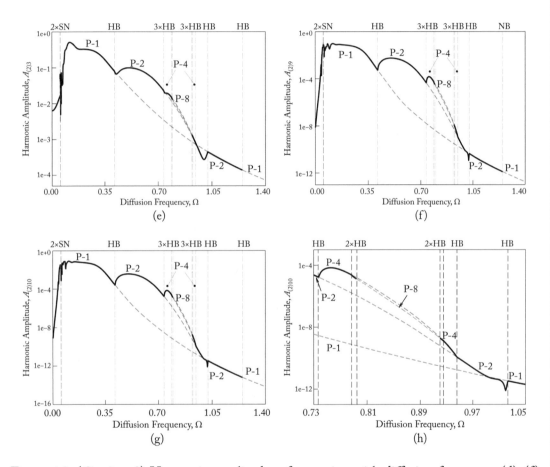

Figure 4.3: (*Continued.*) Harmonic amplitudes of y varying with diffusion frequency: (d)–(f) harmonic amplitude $A_{(2)2}$, $A_{(2)3}$, $A_{(2)9}$; (g)–(h) harmonic amplitude $A_{(2)10}$. Parameter ($a = 0.4$, $b = 1.2$, $Q_0 = 0.08$).

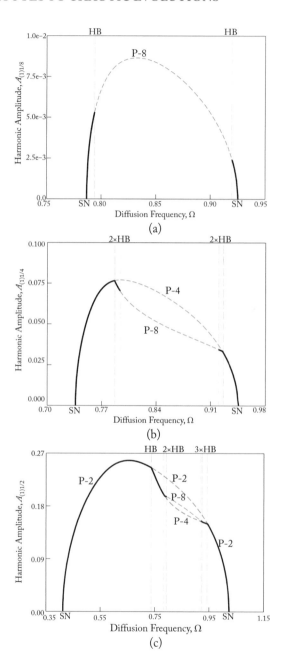

Figure 4.4: Harmonic amplitudes of x varying with diffusion frequency: (a)–(c) harmonic amplitude $A_{(1)k/m}$ ($k = 1, 2, 4$; $m = 8$). Parameter ($a = 0.4$, $b = 1.2$, $Q_0 = 0.08$). (*Continues.*)

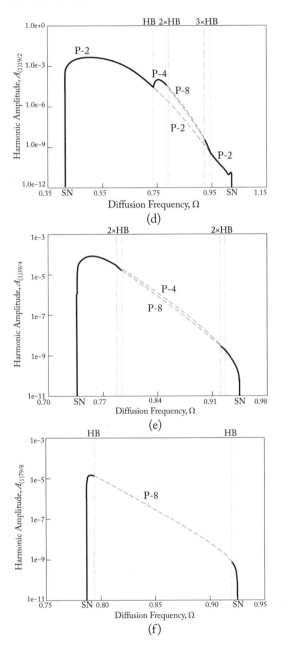

Figure 4.4: (*Continued.*) Harmonic amplitudes of *x* varying with diffusion frequency: (d)–(f) harmonic amplitude $A_{(1)k/m}$ ($k = 76, 78, 79$; $m = 8$). Parameter ($a = 0.4$, $b = 1.2$, $Q_0 = 0.08$).

Table 4.2: Input data for numerical illustrations ($a = 0.4$, $b = 1.2$, $Q_0 = 0.08$)

	Ω	Initial Conditions	P-m Evolution
Figure 4.6	0.3	$(x_0, y_0) \approx (0.993866, 0.961325)$	P-1 (stable)
Figure 4.7	0.7	$(x_0, y_0) \approx (0.281649, 2.285145)$	P-2 (stable)
Figure 4.8	0.75	$(x_0, y_0) \approx (0.291130, 2.157521)$	P-4 (stable)
Figure 4.9	0.79	$(x_0, y_0) \approx (0.283251, 2.179638)$	P-8 (stable)

For comparison, the harmonic amplitudes of three fractional orders ($A_{(2)1/8}$, $A_{(2)1/4}$, $A_{(2)1/2}$) for $k/m \in (0, 1)$ and three fractional orders ($A_{(2)19/2}$, $A_{(2)39/4}$, $A_{(2)79/8}$) for $k/m \in (9, 10)$ are presented in Fig. 4.5(a)–(f) for the concentration y. The bifurcation trees based on such harmonic amplitudes are clearly presented and the quantity levels of harmonic amplitudes for concentration y are similar to the concentration x. The maximum quantity level of $A_{(2)1/2}$ is about $A_{(2)1/2} \sim 0.8$ compared to $A_{(1)1/2} \sim 0.26$; the maximum quantity level of $A_{(2)1/4}$ is $A_{(2)1/4} \sim 0.4$ but $A_{(1)1/4} \sim 0.075$. For period-8 evolutions, $A_{(1)1/8} \sim 9 \times 10^{-3}$ and $A_{(2)1/8} \sim 9 \times 10^{-2}$.

4.2   PERIODIC EVOLUTIONS ON BIFURCATION TREE

In this section, numerical illustrations based on the analytical solutions and numerical integration schemes are presented. The initial conditions in numerical simulations are obtained from approximate analytical solutions of periodic solutions in the previous section. Input data comprising system parameters and initial conditions for numerical simulations are tabulated in Table 4.2. In all plots for illustration, circular symbols gives approximate solutions, and solid curves give numerical simulation results. The acronym "I.C." with a large circular symbol represents initial condition for all plots. The numerical solutions of periodic motions are generated via the mid-point scheme.

Unstable periodic evolutions can hardly be simulated by numerical method due to accumulative error. Therefore, only stable period-m ($m = 1, 2, 4, 8$) evolutions are obtained herein to illustrate the analytical solution in the previous section. The initial conditions for numerical simulations are obtained from the analytical solutions by setting $t = 0$ in Eq. (3.10). Parameters and initial conditions for period-m ($m = 1, 2, 4, 8$) evolutions are given in Table 4.2.

To illustrate the chemical reaction dynamics, the concentration orbits (x, y) for periodic evolutions on the bifurcation tree is presented in Fig. 4.6. In Fig. 4.6(a), the orbit of the two concentrations is presented with an alike triangular closed curve rather than a circular closed curve, which cannot be obtained from the traditional perturbation analysis. This periodic concentration orbit possesses a slow-varying portion and a fast-changing portion. The slow-varying portion is related to the small concentration rate. For $\Omega = 0.7$ with $(x_0, y_0) \approx (0.281649, 2.285145)$,

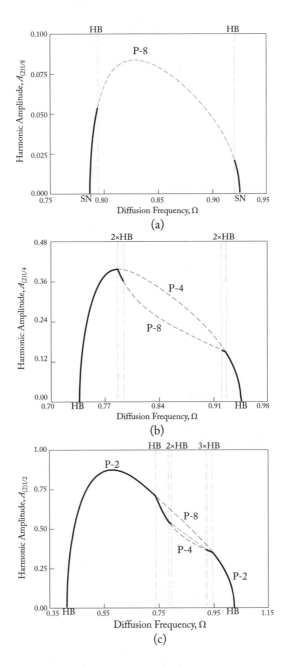

Figure 4.5: Harmonic amplitudes of x varying with diffusion frequency: (a)–(c) harmonic amplitude $A_{(2)k/m}$ ($k = 1, 2, 4$; $m = 8$). Parameter ($a = 0.4$, $b = 1.2$, $Q_0 = 0.08$). (*Continues.*)

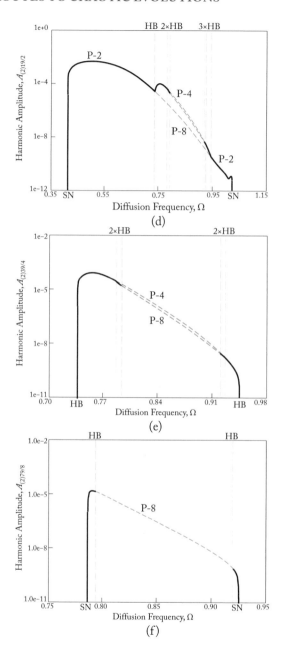

Figure 4.5: (*Continued.*) Harmonic amplitudes of x varying with diffusion frequency: (d)–(f) harmonic amplitude $A_{(2)k/m}$ ($k = 76, 78, 79$; $m = 8$). Parameter ($a = 0.4$, $b = 1.2$, $Q_0 = 0.08$).

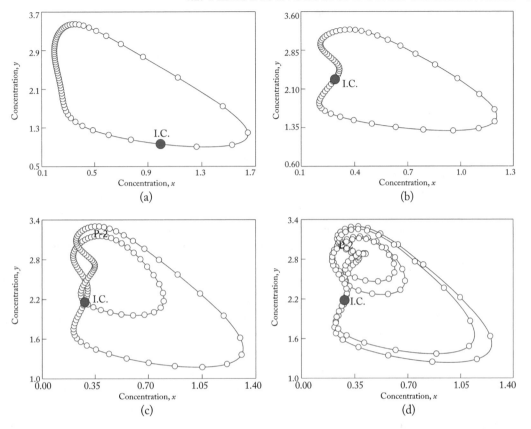

Figure 4.6: Concentration orbit (x, y) for periodic evolutions on the bifurcation tree: (a) period-1 evolution ($\Omega = 0.3$, $x_0 \approx 0.993866$, $y_0 \approx 0.961325$); (b) period-2 evolution ($\Omega = 0.7$, $x_0 \approx 0.291130$, $y_0 \approx 2.285145$); (c) period-4 evolution ($\Omega = 0.75$); (d) period-8 evolution ($\Omega = 0.79$). Parameters ($a = 0.4$, $b = 1.2$, $Q_0 = 0.08$).

the orbit of the two concentrations for period-2 evolutions is presented in Fig. 4.6(b). Such a period-2 orbit still has one cycle orbit. The period-2 concentration orbit still possesses a slow-varying portion and a fast-varying portion. The slow-varying portion has been deformed for comparison to the period-1 concentration orbit. In Fig. 4.6(c), the orbit of the two concentrations for a period-4 evolution with two closed cycles is presented for $\Omega = 0.75$ with $(x_0, y_0) \approx (0.291130, 2.157521)$. The period-4 concentration orbit still possesses two slow-varying portions and two fast-changing portions. The two slow-varying portions have been doubled for the period-2 concentration orbit. The two fast-changing portions are different. One fast-changing portion is much slower than another fast-changing portion. In Fig. 4.6(d), the orbit of the two concentrations for a period-8 evolution with four closed cycles is presented for $\Omega = 0.79$ with $(x_0, y_0) \approx (0.283251, 2.179638)$, which double the period-4 evolution. The period-8 concentration orbit still possesses four (4) slow-varying portions and four (4) fast-changing portions. The two intermediate fast-changing portions are similar and the two large fast-changing portions are similar.

4.2.1   PERIOD-1 EVOLUTION

In Fig. 4.7, the period-1 evolution of the Brusselator is presented for $\Omega = 0.3$ with $(x_0, y_0) \approx (0.993866, 0.961325)$. The fast-changing portion has a big concentration rate variation. The trajectories of the concentrations x and y with the corresponding rates $\dot{x} = dx/dt$ and $\dot{y} = dy/dt$ are presented in Fig. 4.7(a) and (b), respectively. For the trajectory of $(x, \dot{x})$, the slow-varying portion has the concentration change rate be almost zero, and the fast-varying portion experiences the large variation of the concentration rate. For the trajectory of $(y, \dot{y})$, the slow-varying portion has an almost constant change rate, and the fast-varying portion has a parabolic negative velocity variation.

The time histories of the two concentrations x and y are plotted in Fig. 4.7(c) and (d), respectively. The concentration x for the period-1 motion has an almost constant concentration with a spike in its time history. Such a phenomenon can be found in many bio-oscillators. The concentration y has an asymmetric parabolic curve in its time history. The two concentrations are positive, which means reactants exist. So they cannot be negative.

For a further understanding of the periodic evolution of the Brusselator, the time histories of two concentration rates are presented in Fig. 4.7(e) and (f), respectively. In Fig. 4.7(e), the concentration rate $\dot{x}$ is almost zero with a fast switching from a positive maximum $(+1.0)$ to negative minimum value (-0.8). The almost zero rate of the concentration goes to the maximum value with a rapid variation, after rapid switching, the negative minimum concentration rate fast returns back to the almost zero concentration rate. For the rapid changes and switching, there is a big chemical reaction force with direction changes for the concentration x. In Fig. 4.6(f), the concentration rate $\dot{y}$ are almost constant with a negative spike. For such a spike, the concentration y fast decreases to a certain level, and then with an almost constant reaction rate to increase the concentration to the maximum value to repeat such a chemical reaction process.

To understand the period-1 evolution, the harmonic amplitudes are very important. The harmonic spectrums of the concentrations x and y are presented in Fig. 4.6(g) and (h), respectively. For the concentration x, the constant $a_{(1)0} = 0.4$. The main harmonic amplitudes are $A_{(1)1} \approx 0.3296$, $A_{(1)2} \approx 0.2289$, $A_{(1)3} \approx 0.1746$, $A_{(1)4} \approx 0.1303$, $A_{(1)5} \approx 0.0965$, $A_{(1)6} \approx 0.0712$, $A_{(1)7} \approx 0.0527$, $A_{(1)8} \approx 0.0392$, $A_{(1)9} \approx 0.0294$, $A_{(1)10} \approx 0.0222$, $A_{(1)11} \approx 0.0170$, $A_{(1)12} \approx 0.0130$, $A_{(1)13} \approx 0.0101$. The other harmonic amplitudes are $A_{(1)k} \in (10^{-14}, 10^{-3})$ for $k = 14, 15, \ldots, 140$ with $A_{(1)140} \approx 1.41 \times 10^{-14}$. For the concentration y, the constant $a_{(2)0} \approx 2.4842$. The main harmonic amplitudes are $A_{(2)1} \approx 0.9375$, $A_{(2)2} \approx 0.4449$, $A_{(2)3} \approx 0.2610$, $A_{(2)4} \approx 0.1696$, $A_{(2)5} \approx 0.1160$, $A_{(2)6} \approx 0.0815$, $A_{(2)7} \approx 0.0584$, $A_{(2)8} \approx 0.0425$, $A_{(2)9} \approx 0.0313$, $A_{(2)10} \approx 0.0234$, $A_{(2)11} \approx 0.0177$, $A_{(2)12} \approx 0.0135$, $A_{(2)13} \approx 0.0104$. The other har-

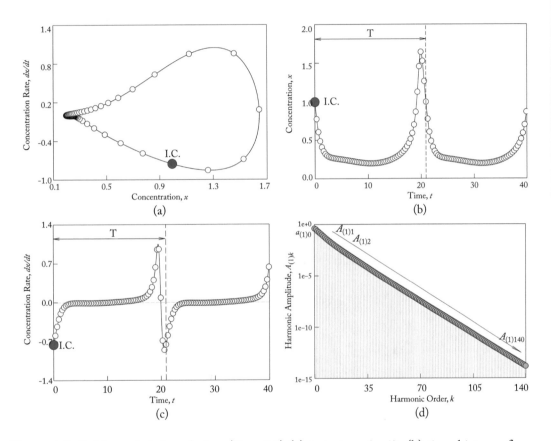

Figure 4.7: Stable period-1 evolution ($\Omega = 0.3$) (a) trajectory $(x, \dot{x})$; (b) time history of concentration (t, x); (c) time history of concentration rate $(t, \dot{x})$; (d) harmonic amplitude spectrum of concentration x. Parameters ($a = 0.4$, $b = 1.2$, $Q_0 = 0.08$), ($x_0 \approx 0.993866$ and $y_0 \approx 0.961325$). (*Continues.*)

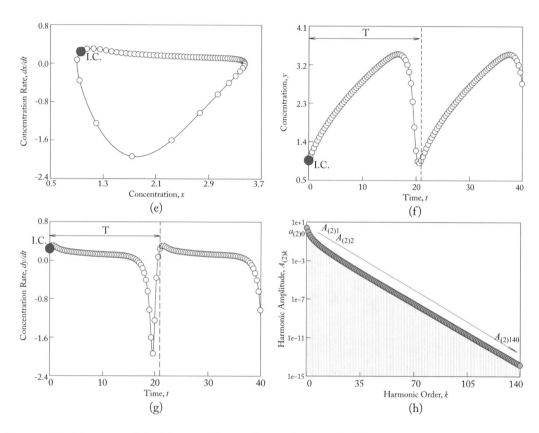

Figure 4.7: (*Continued.*) Stable period-1 evolution ($\Omega = 0.3$) (e) trajectory ($y, \dot{y}$); (f) time history of concentration (t, y); (g) time history of concentration rate ($t, \dot{y}$); (h) harmonic amplitude spectrum of concentration y. Parameters ($a = 0.4$, $b = 1.2$, $Q_0 = 0.08$), ($x_0 \approx 0.993866$ and $y_0 \approx 0.961325$).

monic terms are $A_{(2)k} \in (10^{-14}, 10^{-3})$ for $k = 14, 15, \ldots, 140$ with $A_{(1)140} \approx 1.41 \times 10^{-14}$. For this period-1 evolution, the harmonic amplitude varying with harmonic order is very slow. The 13th-order harmonic amplitude is quantity level of 10^{-3}. Thus, such a period-1 evolution needs at least 13 harmonic terms to get a reasonably accurate solution with the error less than 10^{-3}. Traditional method with one or two harmonic terms cannot provide the adequate solution, which cannot be used for guiding experiment. From the harmonic spectrum, the analytical period-1 evolution is about the accuracy of 10^{-14} with 140 terms. For mechanical oscillators, just a few harmonic terms can be achieve such accuracy of 10^{-14}. In addition, such a period-1 evolution in the Brusselator possesses the characteristics of the van der Pol oscillator, which is difficult to the periodic motion. The period-1 evolution possesses the slow-varying portion and the fast-varying portion. The fast-varying portion forms the spike phenomenon.

On the left side of the bifurcation tree, at $\Omega \approx 0.413493$, the period-1 evolution has a Hopf bifurcation, and the stable peiod-1 evolution becomes unstable. The period-2 evolution appears. At $\Omega \approx 0.736829$, the period-2 evolution has a Hopf bifurcation. The stable period-2 evolution becomes unstable and a period-4 evolution appears. On the right side of bifurcation tree, at $\Omega \approx 1.023424$, the period-1 evolution has a Hopf bifurcation, and the stable period-1 evolution becomes unstable and period-2 evolution appears. With decreasing diffusion frequency, the period-2 evolution has a Hopf bifurcation at $\Omega \approx 0.945221$. The stable period-2 evolution becomes unstable and the period-4 evolution appears. Thus, the stable period-2 evolution is in the two ranges of $\Omega \in (0.945221, 1.023424)$ and $(0.413493, 0.736829)$.

4.2.2   PERIOD-2 EVOLUTION

In Fig. 4.8, the period-2 evolution of the Brusselator is presented first for $\Omega = 0.7$ with $(x_0, y_0) \approx (0.281649, 2.285145)$. The fast-changing portion did not change too much. The trajectories of the concentrations x and y with the corresponding rates are presented in Fig. 4.8(a) and (c), respectively. For the trajectory of $(x, \dot{x})$, the slow-varying portion has a small cycle, and the fast-varying portion still look like the period-1 evolution. For the trajectory of $(y, \dot{y})$, the slow-varying portion has a small wavy change rate, and the fast-varying portion still look like a parabolic negative velocity variation.

The time histories of the two concentrations x and y are plotted in Fig. 4.8(b) and (d), respectively. The concentration x for the period-2 motion has a small wavy concentration with a spike in its time history. The concentration y has an asymmetric parabolic curve in its time history, and just the location of maximum concentration value has been changed. To know the slow-varying concentration waving, the time histories of two concentration rates are presented in Fig. 4.8(e) and (g), respectively. In Fig. 4.8(f), the concentration rate $\dot{x}$ is an alike small sine wave with a fast large switching from a positive maximum ($+0.5$) to negative minimum value (-0.5). In Fig. 4.8(h), the concentration rate $\dot{y}$ are almost constant with a negative spike, which is similar to the period-1 evolution.

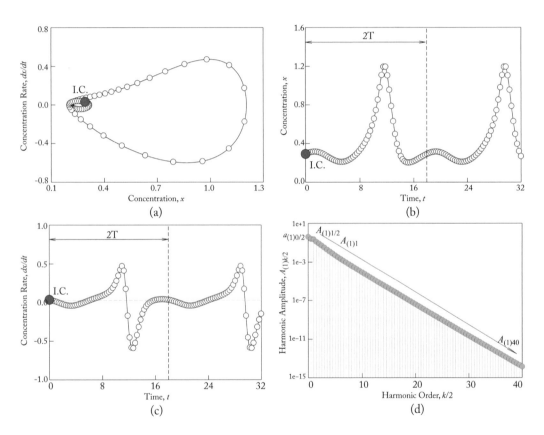

Figure 4.8: Stable period-2 evolution ($\Omega = 0.7$): (a) trajectory $(x, \dot{x})$; (b) time history of concentration (t, x); (c) time history of concentration rate $(t, \dot{x})$; (d) harmonic amplitude spectrum of concentration x. Parameters ($a = 0.4$, $b = 1.2$, $Q_0 = 0.08$), ($x_0 \approx 0.291130$ and $y_0 \approx 2.285145$). (*Continues.*)

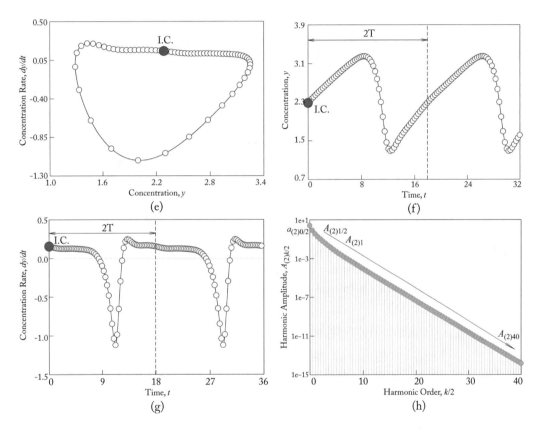

Figure 4.8: (*Continued.*) Stable period-2 evolution ($\Omega = 0.7$): (e) trajectory ($y, \dot{y}$); (f) time history of concentration (t, y); (g) time history of concentration rate ($t, \dot{y}$); (h) harmonic amplitude spectrum of concentration y. Parameters ($a = 0.4$, $b = 1.2$, $Q_0 = 0.08$), ($x_0 \approx 0.291130$ and $y_0 \approx 2.285145$).

To understand the period-2 evolution, the harmonic spectrums of the concentrations x and y are presented in Fig.7(viii) and (ix), respectively. For the concentration x, the constant $a_{(1)0/2} = 0.4$. The main harmonic amplitudes are $A_{(1)1/2} \approx 0.2538$, $A_{(1)1} \approx 0.2253$, $A_{(1)3/2} \approx 0.1250$, $A_{(1)5/2} \approx 0.0511$, $A_{(1)3} \approx 0.0319$, $A_{(1)7/2} \approx 0.0196$, $A_{(1)4} \approx 0.0122$. The other harmonic terms have $A_{(1)k/2} \in (10^{-14}, 10^{-3})$ for $k = 9, 10, \ldots, 80$ with $A_{(1)40} \approx 1.37 \times 10^{-14}$. For the concentration y, the constant $a_{(2)0/2} \approx 2.4366$. The main harmonic amplitudes are $A_{(2)1/2} \approx 0.7682$, $A_{(2)1} \approx 0.3381$, $A_{(2)3/2} \approx 0.1726$, $A_{(2)2} \approx 0.1013$, $A_{(2)5/2} \approx 0.0589$, $A_{(2)3} \approx 0.0353$, $A_{(2)7/2} \approx 0.0212$, $A_{(2)4} \approx 0.0129$. The other harmonic terms are $A_{(2)k/2} \in (10^{-14}, 10^{-3})$ for $k = 9, 10, \ldots, 80$ with $A_{(2)40} \approx 1.37 \times 10^{-14}$. For this period-2 evolution, the harmonic amplitude varying with harmonic order is very slow. The 8th-order harmonic amplitude is quantity level of 10^{-3}. Thus, such a period-2 evolution needs at least eight (8) harmonic terms to get a reasonably accurate solution with the error less than 10^{-3}. From the harmonic spectrum, the analytical period-2 evolution is about the accuracy of 10^{-14} with 80 terms.

On the left side of the bifurcation tree, at $\Omega \approx 0.736829$, the period-2 evolution has a Hopf bifurcation. The stable peiod-2 evolution becomes unstable, and the period-4 evolution appears. At $\Omega \approx 0.78681$, the period-4 evolution has a Hopf bifurcation. The stable period-4 evolution becomes unstable and a period-8 evolution appears. On the right side of bifurcation tree, at $\Omega \approx 0.945221$, the period-2 evolution has a Hopf bifurcation, and the stable period-2 evolution becomes unstable and period-4 evolution appears. With diffusion frequency decrease, the period-4 evolution has a Hopf bifurcation at $\Omega \approx 0.92512$. The stable period-4 evolution becomes unstable and the period-4 evolution appears. Thus, the stable period-4 evolution is in $\Omega \in (0.92512, 0.945221)$ and $(0.736829, 0.78681)$.

4.2.3   PERIOD-4 EVOLUTION

In Fig. 4.9, the period-4 evolution of the Brusselator is presented for $\Omega = 0.75$ with $(x_0, y_0) \approx (0.291130, 2.157521)$. The trajectories of the concentrations x and y with the corresponding rates are presented in Fig. 4.9(a) and (b), respectively. For the trajectory of $(x, \dot{x})$, the two slow-varying portions forms two small cycles, and the two fast-varying portion forms two large cycles. For the trajectory of $(y, \dot{y})$, there are two closed cycles with two slow-varying portions and two fast portions. The time histories of the two concentrations x and y are plotted in Fig. 4.9(c) and (d), respectively. The concentration x for the period-4 motion has two small, wavy concentrations with two spikes in the time history. The concentration y has two asymmetric parabolic curves in the time history. The time histories of two concentration rates are presented in Fig. 4.9(e) and (f), respectively. In Fig. 4.9(e), the concentration rate $\dot{x}$ has two small sine waves and two fast large switching waves. In Fig. 4.9(f), the concentration rate $\dot{y}$ has two small waving constants with two different negative spikes.

To understand the period-4 evolution, the harmonic spectrums of the concentrations x and y are presented in Fig. 4.8(g) and (h), respectively. For the concentration x, the constant $a_{(1)0/4} = 0.4$. The main harmonic amplitudes are $A_{(1)1/4} \approx 0.0502$, $A_{(1)1/2} \approx 0.2316$, $A_{(1)3/4} \approx$

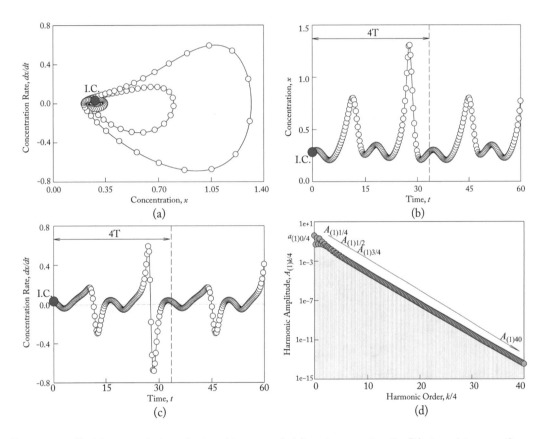

Figure 4.9: Stable period-4 evolution ($\Omega = 0.75$) (a) trajectory $(x, \dot{x})$; (b) time history of concentration (t, x); (c) time history of concentration rate $(t, \dot{x})$; (d) harmonic amplitude spectrum of concentration x. Parameters ($a = 0.4$, $b = 1.2$, $Q_0 = 0.08$), ($x_0 \approx 0.281649$ and $y_0 \approx 2.157521$). (*Continues.*)

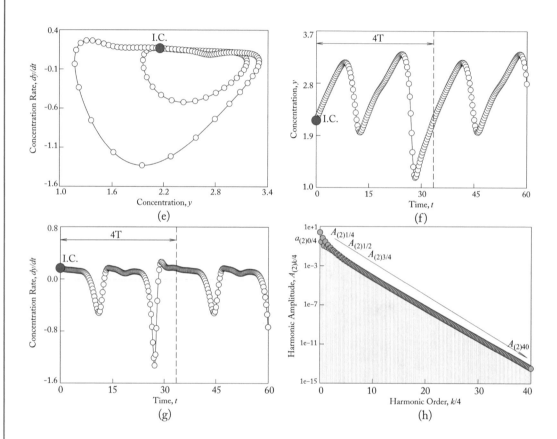

Figure 4.9: (*Continued.*) Stable period-4 evolution ($\Omega = 0.75$) (e) trajectory ($y, \dot{y}$); (f) time history of concentration (t, y); (g) time history of concentration rate ($t, \dot{y}$); (h) harmonic amplitude spectrum of concentration y. Parameters ($a = 0.4$, $b = 1.2$, $Q_0 = 0.08$), ($x_0 \approx 0.281649$ and $y_0 \approx 2.157521$).

0.0541, $A_{(1)1} \approx 0.1916$, $A_{(1)5/4} \approx 0.0547$, $A_{(1)3/2} \approx 0.0941$, $A_{(1)7/4} \approx 0.0453$, $A_{(1)2} \approx 0.0556$, $A_{(1)9/4} \approx 0.0336$, $A_{(1)5/2} \approx 0.0307$, $A_{(1)11/4} \approx 0.0232$, $A_{(1)3} \approx 0.0177$, $A_{(1)7/2} \approx 0.0104$. The other harmonic terms $A_{(1)k/4} \in (10^{-14}, 10^{-3})$ for $k = 15, 16, \ldots, 160$ with $A_{(1)44} \approx 3.39 \times 10^{-14}$. For the concentration y, the constant $a_{(2)0/4} \approx 2.5220$. The main harmonic amplitudes are $A_{(2)1/4} \approx 0.2723$, $A_{(2)1/2} \approx 0.6596$, $A_{(2)3/4} \approx 0.1104$, $A_{(2)1} \approx 0.2808$, $A_{(2)5/4} \approx 0.0800$, $A_{(2)3/2} \approx 0.1259$, $A_{(2)7/4} \approx 0.0569$, $A_{(2)2} \approx 0.0668$, $A_{(2)9/4} \approx 0.0391$, $A_{(2)5/2} \approx 0.0347$, $A_{(2)11/4} \approx 0.0258$, $A_{(2)3} \approx 0.0193$, $A_{(2)13/4} \approx 0.0164$, $A_{(2)7/2} \approx 0.0111$, $A_{(2)15/4} \approx 0.0103$. The other harmonic terms are $A_{(2)k/4} \in (10^{-14}, 10^{-3})$ for $k = 16, 17, \ldots, 160$ with $A_{(2)40} \approx 3.39 \times 10^{-14}$. For this period-3 evolution, the harmonic amplitude varying with harmonic order is very slow. The $(15/4)$th-order harmonic amplitude is quantity level of 10^{-3}. Thus, such a period-4 evolution needs at least 15 harmonic terms to get a reasonably accurate solution with the error less than 10^{-3}. From the harmonic spectrum, the analytical period-4 evolution is about the accuracy of 10^{-14} with 160 terms.

On the left side of the bifurcation tree, at $\Omega \approx 0.78681$, the period-4 evolution has a Hopf bifurcation. The stable peiod-4 evolution becomes unstable, and the period-8 evolution appears. At $\Omega \approx 0.7947$, the period-4 evolution has a Hopf bifurcation. The stable period-4 evolution becomes unstable and a period-8 evolution appears. On the right side of bifurcation tree, at $\Omega \approx 0.92512$, the period-4 evolution has a Hopf bifurcation, and the stable period-4 evolution becomes unstable and period-8 evolution appears. With diffusion frequency decrease, the period-4 evolution has a Hopf bifurcation at $\Omega \approx 0.9199$. The stable period-4 evolution becomes unstable and the period-4 evolution appears. Thus, the stable period-4 evolution is in $\Omega \in (0.91990, 0.92512)$ and $(0.78681, 0.79470)$.

4.2.4   PERIOD-8 EVOLUTION

In Fig. 4.10, the period-8 evolution of the Brusselator is presented for $\Omega = 0.79$ with $(x_0, y_0) \approx (0.283251, 2.179638)$. The trajectories of the concentrations x and y with the corresponding rates are presented in Fig. 4.10(a) and (b), respectively. For the trajectory of $(x, \dot{x})$, the four slow-varying portions forms four small cycles, and the two intermediate-fast and large-fast-varying portions form two relatively smaller cycles and two large cycles. For the trajectory of $(y, \dot{y})$, there are four closed cycles with fours slow-varying portions and four fast portions. The time histories of the two concentrations x and y are plotted in Fig. 4.10(c) and (d), respectively. The concentration x for the period-8 motion has four small, wavy concentration with two small spikes and two big spikes in the time history. The concentration y has four deformed parabolic curves in the time history. The time histories of two concentration rates are presented in Fig. 4.10(e) and (f), respectively. In Fig. 4.10(e), the concentration rate $\dot{x}$ has four small sine waves, two intermediate-fast-switching waves, and two fast-large-switching waves. In Fig. 4.10(f), the concentration rate $\dot{y}$ has four small-waving constants with two large-negative spikes and two intermediate-large negative spikes.

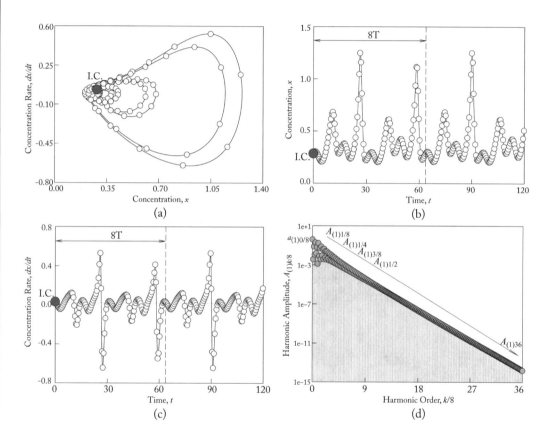

Figure 4.10: Stable period-8 evolution ($\Omega = 0.79$) (a) trajectory $(x, \dot{x})$; (b) time history of concentration (t, x); (c) time history of concentration rate $(t, \dot{x})$; (d) harmonic amplitude spectrum of concentration x. Parameters ($a = 0.4$, $b = 1.2$, $Q_0 = 0.08$), ($x_0 \approx 0.283251$ and $y_0 \approx 2.179638$). (*Continues.*)

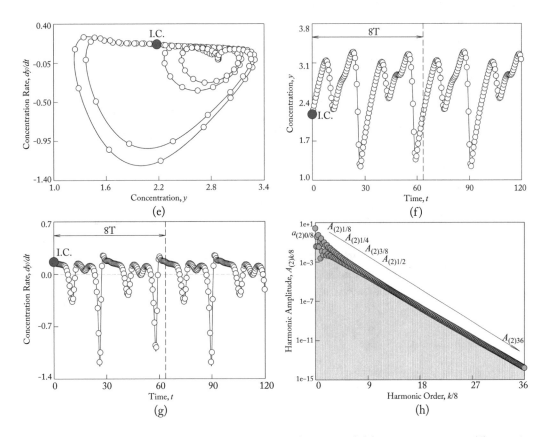

Figure 4.11: (*Continued.*) Stable period-8 evolution ($\Omega = 0.79$) (e) trajectory ($y, \dot{y}$); (f) time history of concentration (t, y); (g) time history of concentration rate ($t, \dot{y}$); (h) harmonic amplitude spectrum of concentration y. Parameters ($a = 0.4$, $b = 1.2$, $Q_0 = 0.08$), ($x_0 \approx 0.283251$ and $y_0 \approx 2.179638$).

For the period-8 evolution, the harmonic spectrums of the concentrations x and y are presented in Fig. 4.10(g) and (viii), respectively. For the concentration x, the constant $a_{(1)0/8} = 0.4$. The main harmonic amplitudes are $A_{(1)1/8} \approx 3.7266e - 3$, $A_{(1)1/4} \approx 0.0736$, $A_{(1)3/8} \approx 0.0111$, $A_{(1)1/2} \approx 0.1971$, $A_{(1)5/8} \approx 0.0154$, $A_{(1)3/4} \approx 0.0664$, $A_{(1)7/8} \approx 1.3202e - 3$, $A_{(1)1} \approx 0.1675$, $A_{(1)9/8} \approx 7.8827e - 3$, $A_{(1)5/4} \approx 0.0520$, $A_{(1)11/8} \approx 3.7522e - 3$, $A_{(1)3/2} \approx 0.0737$, $A_{(1)13/8} \approx 8.7950e - 3$, $A_{(1)7/4} \approx 0.0389$, $A_{(1)15/8} \approx 3.9192e - 3$, $A_{(1)2} \approx 0.0431$, $A_{(1)17/8} \approx 5.4099e - 3$, $A_{(1)9/4} \approx 0.0252$, $A_{(1)19/8} \approx 3.5530e - 3$, $A_{(1)5/2} \approx 0.0226$, $A_{(1)21/8} \approx 4.0629e - 3$, $A_{(1)11/4} \approx 0.0159$, $A_{(1)23/8} \approx 2.8694e - 3$, $A_{(1)3} \approx 0.0127$. The other harmonic terms $A_{(1)k/8} \in (10^{-14}, 10^{-3})$ for $k = 25, 26, \ldots, 288$ with $A_{(1)36} \approx 1.54 \times 10^{-14}$. For the concentration y, the constant $a_{(2)0/8} \approx 2.6044$. The main harmonic amplitudes are $A_{(2)1/8} \approx 0.0379$, $A_{(2)1/4} \approx 0.3798$, $A_{(2)3/8} \approx 0.0390$, $A_{(2)1/2} \approx 0.5366$, $A_{(2)5/8} \approx 0.0346$, $A_{(2)3/4} \approx 0.1303$, $A_{(2)7/8} \approx 2.3217e - 3$, $A_{(1)1} \approx 0.2421$, $A_{(2)9/8} \approx 0.0119$, $A_{(1)5/4} \approx 0.0740$, $A_{(2)11/8} \approx 5.1001e - 3$, $A_{(2)3/2} \approx 0.0965$, $A_{(2)13/8} \approx 0.0111$, $A_{(2)7/4} \approx 0.0480$, $A_{(2)15/8} \approx 4.7287e - 3$, $A_{(2)2} \approx 0.0510$, $A_{(2)17/8} \approx 6.2970e - 3$, $A_{(2)9/4} \approx 0.0289$, $A_{(2)19/8} \approx 4.0261e - 3$, $A_{(2)5/2} \approx 0.0253$, $A_{(2)21/8} \approx 4.5106e - 3$, $A_{(2)11/4} \approx 0.0175$, $A_{(2)23/8} \approx 3.1352e - 3$, $A_{(1)3} \approx 0.0138$, $A_{(2)25/8} \approx 2.8344e - 3$, $A_{(2)13/4} \approx 0.0102$. The other harmonic terms are $A_{(2)k/8} \in (10^{-14}, 10^{-3})$ for $k = 27, 28 \ldots, 288$ with $A_{(2)36} \approx 1.54 \times 10^{-14}$. For this period-3 evolution, the harmonic amplitude varying with harmonic order is very slow. The (13/4)th-order harmonic amplitude is quantity level of 10^{-3}. Thus, such a period-8 evolution needs at least 26 harmonic terms to get a reasonably accurate solution with the error less than 10^{-3}. From the harmonic spectrum, the analytical period-4 evolution is about the accuracy of 10^{-14} with 288 harmonic terms.

CHAPTER 5

Independent Periodic Evolutions

In this chapter, the independent periodic evolutions of the periodically forced diffusion Brusselator is presented from Luo and Guo [39].

5.1 FREQUENCY-AMPLITUDE CHARACTERISTICS

The analytical approximate solutions of the Brusselator with a harmonic diffusion are presented through harmonic amplitude vs. diffusion frequency. The acronym "SN" is for saddle-node bifurcation. The stable and unstable periodic evolutions are plotted in solid and dashed curves, respectively.

Consider a set of system parameters as

$$a = 0.4, \ b = 1.2, \ Q_0 = 0.03. \tag{5.1}$$

For convenience, a summary of all independent periodic evolutions in the range of $\Omega \in [0.9, \ 1.2]$ is given in Table 5.1. More details will be discussed later.

The boundaries of diffusion frequency exist at the bifurcation points. With diffusion frequency decrease, period-3 evolution appears at $\Omega_{cr} \approx 1.178726$, which is a saddle-node bifurcation. Such a period-3 evolution disappears at at $\Omega_{cr} \approx 1.088461$ for another saddle-node bifurcation. For period-5 evolution, the upper-frequency boundary is at $\Omega_{cr} \approx 0.978125$, and the lower limit at $\Omega_{cr} \approx 0.959042$. For the period-7 evolution, diffusion frequency is at $\Omega_{cr} \approx 0.924681$ for onset, and $\Omega \approx 0.916663$ for disappearance. The period-9 evolution exists in the range of $\Omega \in (0.899029, 0.901316)$. The interval of frequency range for period-m evolution decreases with periodic order m. It implies that the higher-order periodic evolution might be difficult to obtain because of the narrow existence range of frequency.

5.1.1 PERIOD-3 EVOLUTIONS

The frequency-amplitude characteristics of period-3 evolutions are presented for $\Omega \in (1.088461, 1.178726)$ in Figs. 5.1 and 5.2 for the concentrations x and y, respectively. The frequency-amplitude characteristics of concentration x are presented in Fig. 5.1(a)–(h) for constant $a_{(1)0/3}$ and harmonic amplitude $A_{(1)k/3}$ ($k = 1, 2, 3, 6, 22, 23, 24$). The constant $a_{(1)0/3} = 0.4$ is for all frequency range in Fig. 5.1(a). In Fig. 5.1(b), the harmonic amplitude of $A_{(1)1/3}$ is

Table 5.1: Existence and bifurcation summary for independent period-1 motions ($a = 0.4$, $b = 1.2$, $Q_0 = 0.03$)

	Ω	Simulation
Period-3	(1.088461, 1.178726)	$\Omega = 1.1$
Period-5	(0.959042, 0.978125)	$\Omega = 0.97$
Period-7	(0.916663, 0.924681)	$\Omega = 0.92$
Period-9	(0.899029, 0.901316)	$\Omega = 0.90$

$A_{(1)1/3} \sim 0.13$ at the lower bifurcation $\Omega_{cr} \approx 1.088461$. The quantity level decays to $A_{(1)1/3} \sim 0.08$ in the middle of the range, and $A_{(1)1/3} \sim 0.09$ at $\Omega_{cr} \approx 1.178726$. The harmonic amplitudes of the second terms are presented in Fig. 5.1(c). The harmonic amplitude $A_{(1)2/3}$ changes from 0.01–0.06. The first primary harmonic order is shown in Fig. 5.1(d). The quantity level of harmonic amplitude $A_{(1)3/3}$ is similar to the second fractional order $A_{(1)2/3}$, but the maximum level of such a primary harmonic amplitude is about $A_{(1)3/3} = A_{(1)1} \approx 0.03$. In Fig. 5.1(e), the second primary harmonic term is presented, and the quantity level of the harmonic amplitude is $A_{(1)6/3} = A_{(1)2} < 8 \times 10^{-4}$. The effects of such a harmonic term on period-3 evolution drop dramatically. To avoid abundant illustrations, the harmonic amplitudes of $A_{(1)22/3}$, $A_{(1)23/3}$, and $A_{(1)24/3}$ are presented in Fig. 5.1(f)–(h), respectively. At $\Omega_{cr} \approx 1.088461$, the quantity levels of three harmonic amplitudes are $A_{(1)22/3} \approx 3.6 \times 10^{-11}$, $A_{(1)23/3} \approx 1.3 \times 10^{-11}$, and $A_{(1)24/3} \approx 4.6 \times 10^{-12}$. At $\Omega_{cr} \approx 1.178726$, $A_{(1)22/3} \approx 1.7 \times 10^{-12}$, $A_{(1)23/3} \approx 5.2 \times 10^{-13}$, and $A_{(1)24/3} \approx 1.6 \times 10^{-13}$.

In a similar fashion, the frequency-amplitude characteristics for the concentration y of the period-3 evolution are presented in Fig. 5.2. Two saddle-node bifurcations are located at the upper and lower limits of the existence of such a period-3 evolution. The frequency-amplitude characteristics of concentration y are presented in Fig. 5.2(a)–(h) for constant $a_{(2)0/3}$ and harmonic amplitude $A_{(2)k/3}$ ($k = 1, 2, 3, 6, 22, 23, 24$). In Fig. 5.2(a), the constant term of concentration y is not constant anymore, and the quantity level is for $a_{(2)0/3} \in (2.86, 2.95)$. In Fig. 5.2(b), the harmonic amplitude of $A_{(2)1/3}$ is about $A_{(2)1/3} \sim 0.4$ at $\Omega_{cr} \approx 1.088461$ and $A_{(2)1/3} \sim 0.23$ at $\Omega_{cr} \approx 1.178726$. The harmonic amplitudes of $A_{(2)2/3}$ are presented in Fig. 5.2(c). The harmonic amplitude $A_{(1)2/3}$ changes from 0.01–0.1. The first primary harmonic amplitude is shown in Fig. 5.2(d). Such a primary harmonic amplitude is $A_{(2)3/3} = A_{(2)1} \in (0.001, 0.040)$. For the second primary harmonic term, the frequency-amplitude characteristics is presented in Fig. 5.2(e). The quantity level of the second harmonic amplitude is $A_{(2)6/3} = A_{(2)2} < 0.08$. To avoid abundant illustrations, the harmonic amplitudes of $A_{(2)22/3}$, $A_{(2)23/3}$, and $A_{(2)24/3}$ are presented in Fig. 5.2(f)–(h), respectively. At $\Omega_{cr} \approx 1.088461$, the quantity levels of three harmonic amplitudes are $A_{(1)22/3} \approx 3.7 \times 10^{-11}$, $A_{(1)23/3} \approx 1.24 \times 10^{-11}$, and $A_{(1)24/3} \approx 4.7 \times 10^{-12}$. At $\Omega_{cr} \approx 1.178726$, $A_{(1)22/3} \approx 1.8 \times 10^{-12}$, $A_{(1)23/3} \approx 5.4 \times 10^{-13}$, and $A_{(1)24/3} \approx 1.69 \times 10^{-13}$.

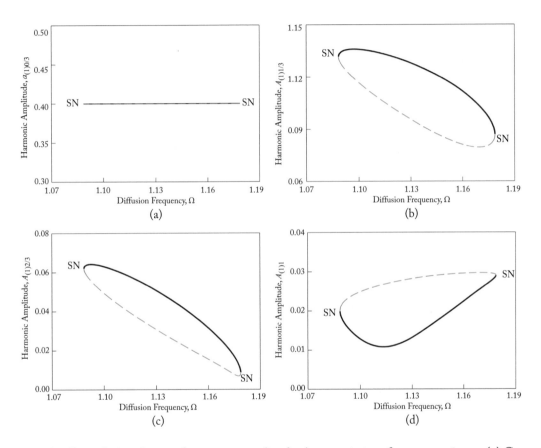

Figure 5.1: Period-3 evolution: frequency-amplitude characteristics of concentration x. (a) Constant $a_{(1)0/3}$; (b)–(h) harmonic amplitude $A_{(1)k/3}$ ($k = 1, 2, 3, 6, 22, 23, 24$), ($a = 0.4$, $b = 1.2$, $Q_0 = 0.03$). The solid and dashed curves are for stable and unstable evolutions, respectively. SN is for saddle-node bifurcation. (*Continues.*)

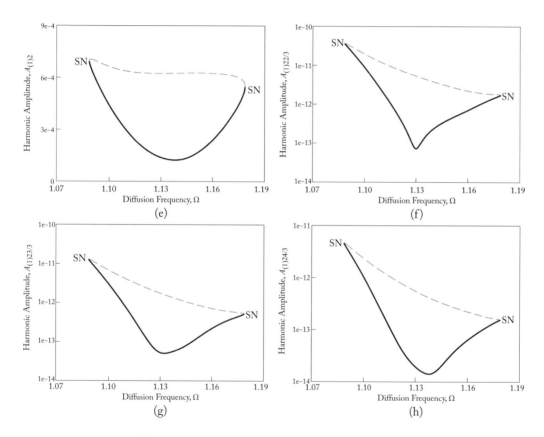

Figure 5.1: (*Continued.*) Period-3 evolution: frequency-amplitude characteristics of concentration x. (b)–(h) harmonic amplitude $A_{(1)k/3}$ ($k = 1, 2, 3, 6, 22, 23, 24$), ($a = 0.4$, $b = 1.2$, $Q_0 = 0.03$). The solid and dashed curves are for stable and unstable evolutions, respectively. SN is for saddle-node bifurcation.

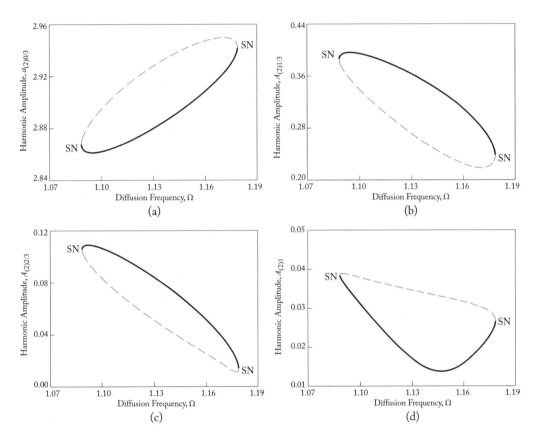

Figure 5.2: Period-3 evolution: frequency-amplitude characteristics of concentration y. (a) constant $a_{(2)0/3}$; (b)–(h) harmonic amplitude $A_{(2)k/3}$ ($k = 1, 2, 3, 6, 22, 23, 24$), ($a = 0.4$, $b = 1.2$, $Q_0 = 0.03$). The solid and dashed curves are for stable and unstable evolutions, respectively. SN is for saddle-node bifurcation. (*Continues.*)

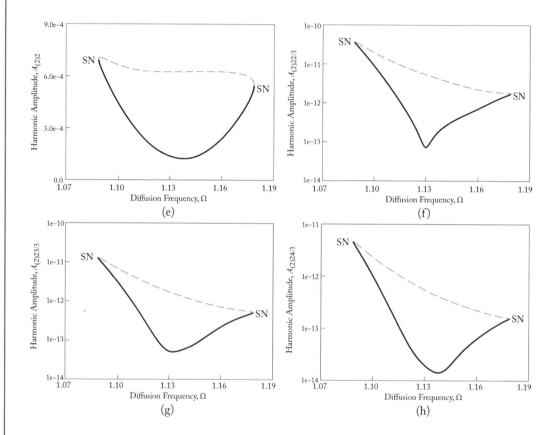

Figure 5.2: (*Continued.*) Period-3 evolution: frequency-amplitude characteristics of concentration y. (b)–(h) harmonic amplitude $A_{(2)k/3}$ ($k = 1, 2, 3, 6, 22, 23, 24$), ($a = 0.4$, $b = 1.2$, $Q_0 = 0.03$). The solid and dashed curves are for stable and unstable evolutions, respectively. SN is for saddle-node bifurcation.

5.1.2   PERIOD-5 EVOLUTIONS

The frequency-amplitude characteristics of period-5 evolutions are presented for $\Omega \in$ (0.959042, 0.978125) in Figs. 5.3 and 5.4 for the concentrations x and y, respectively. The frequency-amplitude characteristics of concentration x are presented in Fig. 5.3(a)–(h) for constant $a_{(1)0/5}$ and harmonic amplitude $A_{(1)k/5}$ ($k = 1, 2, 3, 5, 48, 49, 50$). In Fig. 5.3(a), the constant is constant with $a_{(1)0/5} = 0.4$. Fig. 5.3(b), the harmonic amplitude of $A_{(1)1/5}$ is $A_{(1)1/5} \sim$ 0.013 at $\Omega_{cr} \approx 0.959042$. The quantity level decays to $A_{(1)1/5} \sim 0.007$ at $\Omega_{cr} \approx 0.978125$. The second harmonic amplitude of $A_{(1)2/5} \in (0.088, 0.101)$ is presented in Fig. 5.3(c). The third harmonic term is shown in Fig. 5.3(d), and the corresponding harmonic amplitude is $A_{(1)3/5} \in$ (0.020, 0.026). The first primary harmonic amplitude is presented in Fig. 5.3(e), and the range of harmonic amplitude is $A_{(1)5/5} \in (0.0294, 0.0318)$. To avoid abundant illustrations, the harmonic amplitudes of $A_{(1)48/5}$, $A_{(1)49/5}$, and $A_{(1)50/5}$ are presented in Fig. 5.3(f)–(h), respectively. At $\Omega_{cr} \approx 0.959042$, the quantity levels of three harmonic amplitudes are $A_{(1)48/5} \approx 1.0 \times 10^{-13}$, $A_{(1)49/5} \approx 6.0 \times 10^{-14}$, and $A_{(1)50/5} \approx 2.2 \times 10^{-14}$. At $\Omega_{cr} \approx 0.978125$, $A_{(1)48/5} \approx 4.5 \times 10^{-14}$, $A_{(1)49/5} \approx 2.3 \times 10^{-14}$, and $A_{(1)50/5} \approx 1.1 \times 10^{-13}$.

The frequency-amplitude characteristics of concentration y of the period-5 evolution are presented in Fig. 5.4. Two saddle-node bifurcations are still located at the upper and lower limits of the existence of such a period-5 evolution. The constant $a_{(2)0/5}$ and harmonic amplitude $A_{(2)k/5}$ ($k = 1, 2, 3, 5, 48, 49, 50$) of the concentration y are presented in Fig. 5.4(a)–(h). In Fig. 5.4(a), the constant term of concentration y varies with diffusion frequency, and the quantity level is in the range of $a_{(2)0/5} \in (2.91, 2.94)$. In Fig. 5.4(b), the harmonic amplitude of $A_{(2)1/5}$ is presented with $A_{(2)1/5} \in (0.02, 0.07)$. The harmonic amplitudes of $A_{(2)2/5} \in$ (0.24, 0.28) are presented in Fig. 5.4(c). Compared to the first harmonic term, the second harmonic term plays an important role on period-5 evolution. The third harmonic amplitude is shown in Fig. 5.4(d) with $A_{(2)3/5} \in (0.045, 0.068)$, which has the same quantity level of the first harmonic term. In Fig. 5.4(e), the first primary amplitude of $A_{(2)5/5} \in (0.04, 0.045)$ is presented, and the quantity level is the same as the first-order harmonic term. To avoid abundant illustrations, the harmonic amplitudes of $A_{(2)48/5}$, $A_{(2)49/5}$, and $A_{(2)50/5}$ are presented in Fig. 5.4(f)–(h), respectively. The range of such amplitudes are $A_{(2)48/5} \in (4.3 \times 10^{-14}, 1.2 \times 10^{-13})$, $A_{(2)49/5} \in (2.2 \times 10^{-14}, 6.8 \times 10^{-14})$, and $A_{(2)50/5} \in (4.3 \times 10^{-15}, 3.1 \times 10^{-14})$.

5.1.3   PERIOD-7 EVOLUTIONS

The frequency-amplitude characteristics of period-7 evolutions are presented for $\Omega \in$ (0.916663, 0.924681) in Figs. 5.5 and 5.6 for the concentrations x and y, respectively. The constant $a_{(1)0/7}$ and harmonic amplitude $A_{(1)k/7}$ ($k = 1, 2, 3, 7, 68, 69, 70$) of concentration x are presented in Fig. 5.5(a)–(h). In Fig. 5.5(a), the constant is still constant with $a_{(1)0/7} = 0.4$. In Fig. 5.5(b), the harmonic amplitude of $A_{(1)1/7}$ is $A_{(1)1/7} \sim 7.0 \times 10^{-3}$ at $\Omega_{cr} \approx 0.916663$, and the quantity level decays to $A_{(1)1/7} \sim 5.1 \times 10^{-3}$ at $\Omega_{cr} \approx 0.924681$. The amplitudes of the second harmonic term are presented in Fig. 5.5(c), which are in $A_{(1)2/7} \in (0.0043, 0.0100)$.

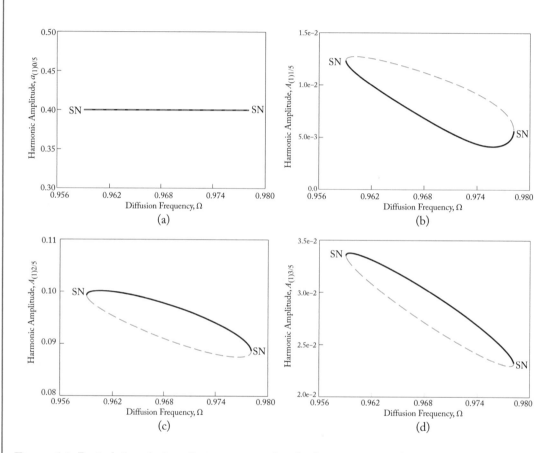

Figure 5.3: Period-5 evolution: frequency-amplitude characteristics of concentration x. (a) constant $a_{(1)0/5}$; (b)–(h) harmonic amplitude $A_{(1)k/5}$ ($k = 1, 2, 3, 5, 48, 49, 50$), ($a = 0.4$, $b = 1.2$, $Q_0 = 0.03$). The solid and dashed curves are for stable and unstable evolutions, respectively. SN is for saddle-node bifurcation. (*Continues.*)

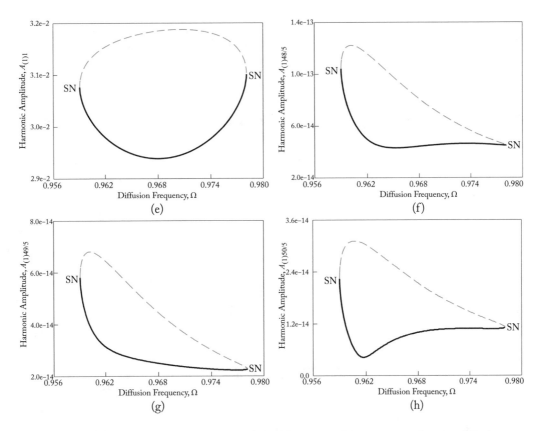

Figure 5.3: (*Continued.*) (b)–(h) harmonic amplitude $A_{(1)k/5}$ ($k = 1, 2, 3, 5, 48, 49, 50$), ($a = 0.4$, $b = 1.2$, $Q_0 = 0.03$). The solid and dashed curves are for stable and unstable evolutions, respectively. SN is for saddle-node bifurcation.

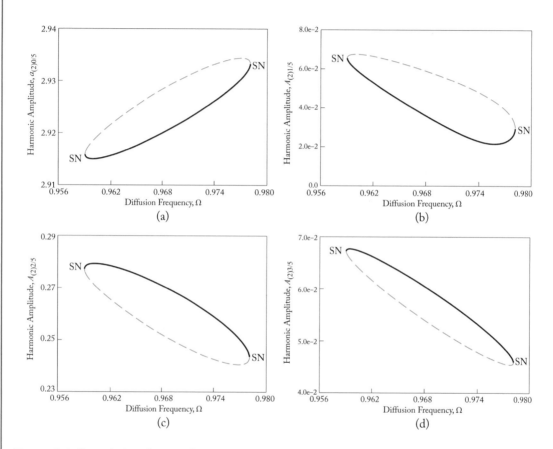

Figure 5.4: Period-5 evolution: frequency-amplitude characteristics of concentration x. (a) constant $a_{(2)0/5}$; (b)–(h) harmonic amplitude $A_{(2)k/5}$ ($k = 1, 2, 3, 5, 48, 49, 50$), ($a = 0.4$, $b = 1.2$, $Q_0 = 0.03$). The solid and dashed curves are for stable and unstable evolutions, respectively. SN is for saddle-node bifurcation. (*Continues.*)

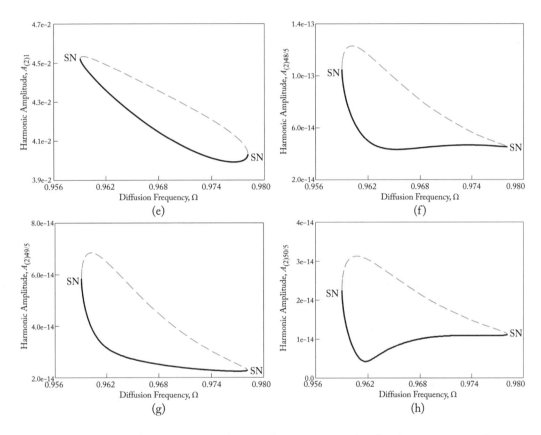

Figure 5.4: (*Continued.*) Period-5 evolution: frequency-amplitude characteristics of concentration x. (b)–(h) harmonic amplitude $A_{(2)k/5}$ ($k = 1, 2, 3, 5, 48, 49, 50$), ($a = 0.4$, $b = 1.2$, $Q_0 = 0.03$). The solid and dashed curves are for stable and unstable evolutions, respectively. SN is for saddle-node bifurcation.

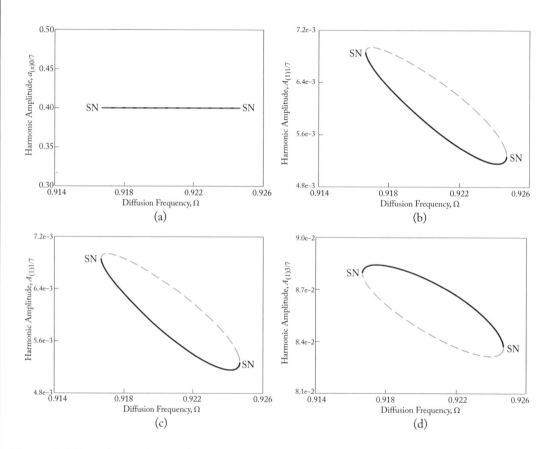

Figure 5.5: Period-7 evolution: frequency-amplitude characteristics of concentration x. (a) constant $a_{(1)0/7}$; (b)–(h) harmonic amplitude $A_{(1)k/7}$ ($k = 1, 2, 3, 7, 68, 69, 70$), ($a = 0.4$, $b = 1.2$, $Q_0 = 0.03$). The solid and dashed curves are for stable and unstable evolutions, respectively. SN is for saddle-node bifurcation. (*Continues.*)

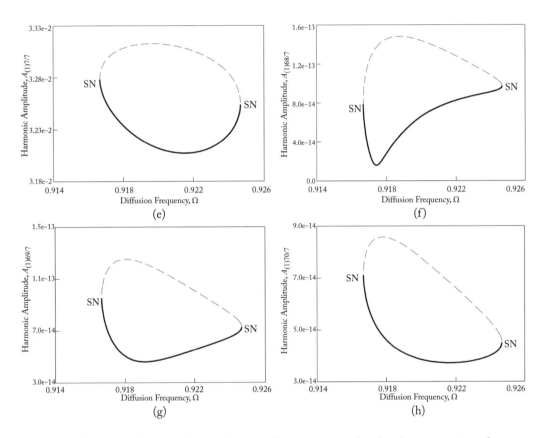

Figure 5.5: (*Continued.*) Period-7 evolution: frequency-amplitude characteristics of concentration x. (b)–(h) harmonic amplitude $A_{(1)k/7}$ ($k = 1, 2, 3, 7, 68, 69, 70$), ($a = 0.4$, $b = 1.2$, $Q_0 = 0.03$). The solid and dashed curves are for stable and unstable evolutions, respectively. SN is for saddle-node bifurcation.

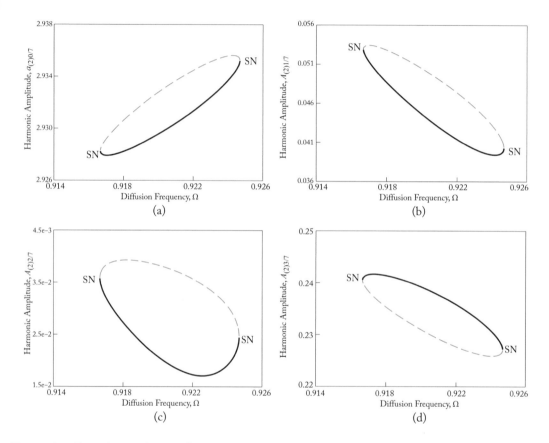

Figure 5.6: Period-7 evolution: frequency-amplitude characteristics of concentration y. (a) constant $a_{(2)0/7}$; (b)–(h) harmonic amplitude $A_{(2)k/7}$ ($k = 1, 2, 3, 7, 68, 69, 70$), ($a = 0.4$, $b = 1.2$, $Q_0 = 0.03$). The solid and dashed curves are for stable and unstable evolutions, respectively. SN is for saddle-node bifurcation. (*Continues.*)

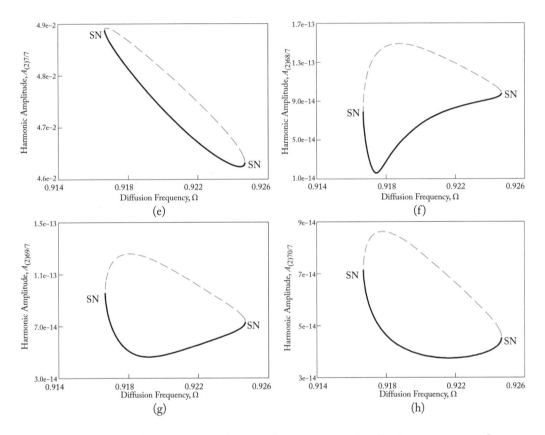

Figure 5.6: (*Continued.*) Period-7 evolution: frequency-amplitude characteristics of concentration y. (b)–(h) harmonic amplitude $A_{(2)k/7}$ ($k = 1, 2, 3, 7, 68, 69, 70$), ($a = 0.4$, $b = 1.2$, $Q_0 = 0.03$). The solid and dashed curves are for stable and unstable evolutions, respectively. SN is for saddle-node bifurcation.

The third harmonic amplitude is $A_{(1)3/7} \in (0.083, 0.088)$, as shown in Fig. 5.5(d). The first primary harmonic amplitude is in $A_{(1)7/7} \in (3.20 \times 10^{-2}, 3.32 \times 10^{-2})$, as shown in Fig. 5.5(e). To avoid abundant illustrations, the harmonic amplitudes of $A_{(1)68/7}, A_{(1)69/7}$, and $A_{(1)70/7}$ are presented in Fig. 5.5(f)–(h), respectively. At $\Omega_{cr} \approx 0.916663$, the quantity levels of three harmonic amplitudes are $A_{(1)68/7} \approx 7.8 \times 10^{-14}$, $A_{(1)69/7} \approx 9.5 \times 10^{-14}$, and $A_{(1)70/7} \approx 7.1 \times 10^{-14}$. At $\Omega_{cr} \approx 0.924681$, $A_{(1)68/7} \approx 9.7 \times 10^{-14}$, $A_{(1)69/7} \approx 7.2 \times 10^{-14}$, and $A_{(1)70/7} \approx 4.5 \times 10^{-14}$.

The frequency-amplitude characteristics of concentration y of the period-7 evolution are presented in Fig. 5.6. The constant $a_{(2)0/7}$ and harmonic amplitude $A_{(2)k/7}$ ($k = 1, 2, 3, 7, 68, 69, 70$) of the concentration y are presented in Fig. 5.6(a)–(h). In Fig. 5.6(a), the constant term of concentration y is in the range of $a_{(2)0/7} \in (2.928, 2.936)$. In Fig. 5.6(b), the harmonic amplitude of $A_{(2)1/7}$ is presented with $A_{(2)1/7} \in (0.040, 0.053)$. The harmonic amplitudes of $A_{(2)2/7} \in (0.017\,0.039)$ are presented in Fig. 5.6(c), which has the same quantity level of the first harmonic term. The third harmonic amplitude is shown in Fig. 5.6(d) with $A_{(2)3/7} \in (0.225, 0.242)$. Compared to the first harmonic term, the second harmonic term plays an important role on period-5 evolution. In Fig. 5.6(e), the first primary amplitude of $A_{(2)7/7} \in (0.046, 0.049)$ is presented, and the quantity level is the same as the first-order harmonic term. To avoid abundant illustrations, the harmonic amplitudes of $A_{(2)68/7}, A_{(2)69/7}$, and $A_{(2)70/7}$ are presented in Fig. 5.6(e)–(h), respectively. The varying range of quantity levels of such higher orders are about $A_{(2)68/7} \in (1.6 \times 10^{-14}, 1.5 \times 10^{-13})$, $A_{(2)69/7} \in (4.6 \times 10^{-14}, 1.3 \times 10^{-13})$, $A_{(2)70/7} \in (3.7 \times 10^{-14}, 8.6 \times 10^{-14})$.

5.1.4 PERIOD-9 EVOLUTIONS

The frequency-amplitude characteristics of period-9 evolutions are presented for $\Omega \in (0.899029, 0.901316)$ in Figs. 5.7 and 5.8 for the concentrations x and y, respectively. The constant $a_{(1)0/9}$ and harmonic amplitude $A_{(1)k/9}$ ($k = 1, 2, 3, 9, 88, 89, 90$) of concentration x are presented in Fig. 5.7(a)–(h). In Fig. 5.7(a), the constant is still constant with $a_{(1)0/9} = 0.4$. The harmonic amplitude of $A_{(1)1/9}$ is $A_{(1)1/9} \sim 4.9 \times 10^{-3}$ at $\Omega_{cr} \approx 0.899029$, and the quantity level decays to $A_{(1)1/9} \sim 4.3 \times 10^{-3}$ at $\Omega_{cr} \approx 0.901316$, as shown in Fig. 5.7(b). The amplitudes of the second harmonic term with $A_{(1)2/9} \in (9 \times 10^{-4}, 4 \times 10^{-3})$ are presented in Fig. 5.7(c). The third harmonic amplitude is $A_{(1)3/9} \in (0.08, 0.011)$, as shown in Fig. 5.7(d). The first primary harmonic amplitude with $A_{(1)9/9} \in (3.32 \times 10^{-2}, 3.38 \times 10^{-2})$ is presented in Fig. 5.7(e). To avoid abundant illustrations, the harmonic amplitudes of $A_{(1)88/9}, A_{(1)89/9}$, and $A_{(1)90/9}$ are presented in Fig. 5.7(f)–(h), respectively. At $\Omega_{cr} \approx 0.899029$, the quantity levels of three harmonic amplitudes are $A_{(1)88/9} \approx 7.7 \times 10^{-14}$, $A_{(1)89/9} \approx 5.4 \times 10^{-14}$, and $A_{(1)90/9} = 6.9 \times 10^{-14}$. At $\Omega_{cr} \approx 0.924681$, $A_{(1)88/9} \approx 1.1 \times 10^{-13}$, $A_{(1)89/9} \approx 8.7 \times 10^{-14}$, and $A_{(1)90/9} = 6.8 \times 10^{-14}$.

The frequency-amplitude characteristics of concentration y of the period-9 evolution are placed in Fig. 5.8. The constant $a_{(2)0/9}$ and harmonic amplitude $A_{(2)k/9}$ ($k = 1, 2, 3, 9, 88, 89, 90$) of the concentration y are presented in Fig. 5.8(a)–(h). In Fig. 5.8(a), the constant term of concentration y is in the range of $a_{(2)0/7} \in (2.9355, 2.9393)$. In Fig. 5.8(b),

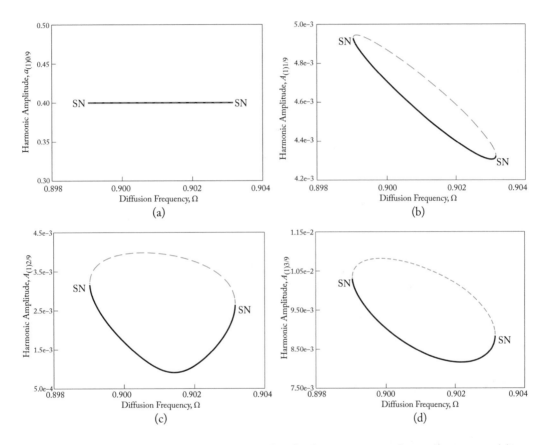

Figure 5.7: Period-9 evolution: frequency-amplitude characteristics of concentration x. (a) constant $a_{(1)0/9}$; (b)–(h) harmonic amplitude $A_{(1)k/9}$ ($k = 1, 2, 3, 7, 88, 89, 90$), ($a = 0.4$, $b = 1.2$, $Q_0 = 0.03$). The solid and dashed curves are for stable and unstable evolutions, respectively. SN is for saddle-node bifurcation. (*Continues.*)

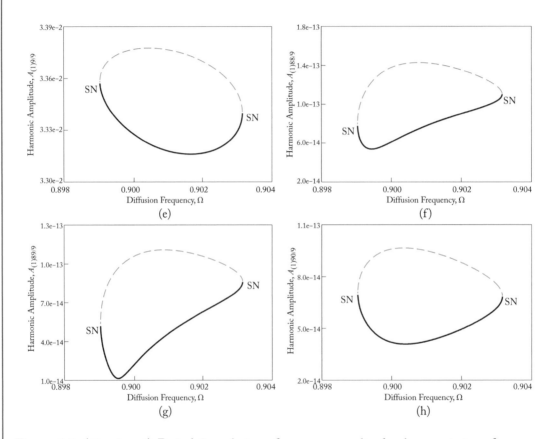

Figure 5.7: (*Continued.*) Period-9 evolution: frequency-amplitude characteristics of concentration x. (b)–(h) harmonic amplitude $A_{(1)k/9}$ ($k = 1, 2, 3, 7, 88, 89, 90$), ($a = 0.4$, $b = 1.2$, $Q_0 = 0.03$). The solid and dashed curves are for stable and unstable evolutions, respectively. SN is for saddle-node bifurcation.

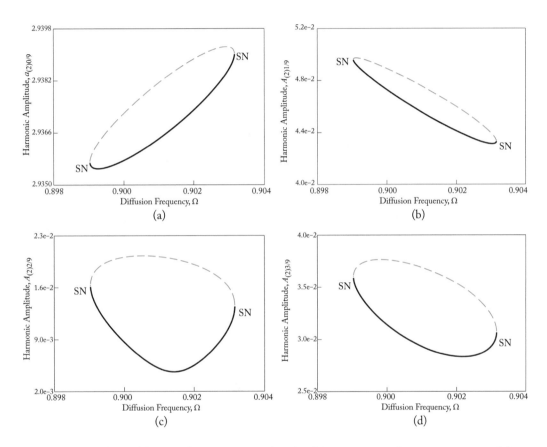

Figure 5.8: Period-9 evolution: frequency-amplitude characteristics of concentration y. (a) constant $a_{(2)0/9}$; (b)–(h) harmonic amplitude $A_{(2)k/9}$ ($k = 1, 2, 3, 7, 88, 89, 90$), ($a = 0.4$, $b = 1.2$, $Q_0 = 0.03$). The solid and dashed curves are for stable and unstable evolutions, respectively. SN is for saddle-node bifurcation. (*Continues.*)

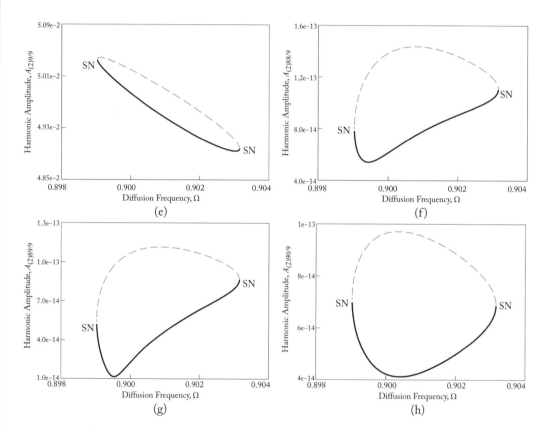

Figure 5.8: (*Continued.*) Period-9 evolution: frequency-amplitude characteristics of concentration y. (b)–(h) harmonic amplitude $A_{(2)k/9}$ ($k = 1, 2, 3, 7, 88, 89, 90$), ($a = 0.4$, $b = 1.2$, $Q_0 = 0.03$). The solid and dashed curves are for stable and unstable evolutions, respectively. SN is for saddle-node bifurcation.

the harmonic amplitude of $A_{(2)1/9}$ is presented with $A_{(2)1/9} \in (0.043, 0.049)$. The harmonic amplitudes of $A_{(2)2/9} \in (0.005, 0.020)$ are presented in Fig. 5.8(c), which has the same quantity level of the first harmonic term. The third harmonic amplitude is shown in Fig. 5.8(d) with $A_{(2)3/9} \in (0.028, 0.038)$. Compared to the first harmonic term, the second harmonic term plays an important role on period-9 evolution. In Fig. 5.8(e), the first primary amplitude of $A_{(2)9/9} \in (0.0490, 0.0504)$ is presented, and the quantity level is the same as the first-order harmonic term. To avoid abundant illustrations, the harmonic amplitudes of $A_{(2)88/9}$, $A_{(2)89/9}$, and $A_{(2)90/9}$ are presented in Fig. 5.8(e)–(h), respectively. The varying range of such higher-order harmonic amplitude of the period-9 evolution are about $A_{(2)88/9} \in (5.4 \times 10^{-14}, 1.4 \times 10^{-13})$, $A_{(2)89/9} \in (1.2 \times 10^{-14}, 1.2 \times 10^{-13})$, and $A_{(2)90/9} \in (4.1 \times 10^{-14}, 9.7 \times 10^{-14})$.

5.2 ILLUSTRATIONS OF INDEPENDENT PERIODIC EVOLUTIONS

In this section, numerical illustrations are presented for the analytical solutions proposed in previous sections. In all plots, the solid and dashed curves are for stable and unstable period-m evolution. For unstable periodic evolutions, numerical simulations will take about 100 periods to move away from the analytical solution and then arrive to the corresponding stable periodic evolution. Evolutions only for two periods are presented. So the unstable periodic evolution cannot be move away to the corresponding stable periodic evolution. In Luo and Guo [18, 19], the comparison of analytical and numerical simulations of periodic evolutions was presented. Herein only numerical results of periodic evolutions are presented.

Herein a pair of stable and unstable periodic evolutions are presented. The initial conditions for numerical simulations are obtained from the analytical solutions by setting $t = 0$ in Eq. (3.10). Parameters and initial conditions for period-m evolutions ($m = 3, 5, 7, 9$) are given in Table 5.2. The mid-point scheme will be adopted herein for numerical simulations. However, harmonic amplitudes from the stable periodic evolutions are presented only.

5.2.1 PERIOD-3 EVOLUTIONS

In Fig. 5.9, the period-3 evolution of the Brusselator is presented first for $\Omega = 1.1$. The paired unstable and stable evolutions in the plane of concentrations (x, y) are clearly presented in Fig. 5.9(a). The orbit of the two concentrations for the stable period-3 evolution is a heart-shaped curve rather than a circular closed curve, which cannot be obtained from the traditional perturbation analysis. Such period-3 concentration orbit possesses a slow-varying portion near the concave part of the heart shape and a fast-changing portion in the remaining segment. The slow-varying portion is related to the small concentration rate. The fast-changing portion has a big concentration rate variation. For a better understanding of the fast and slow-varying orbits, the rate orbit of concentrations are presented in Fig. 5.9(b). The trajectories of the concentrations x and y with the corresponding rates $\dot{x} = dx/dt$ and $\dot{y} = dy/dt$ are presented in Fig. 5.9(c) and

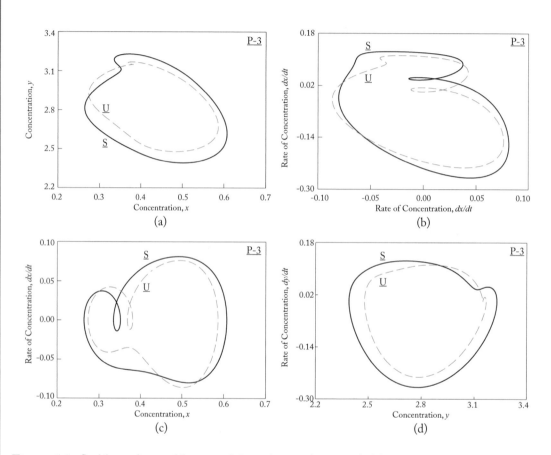

Figure 5.9: Stable and unstable period-3 evolution ($\Omega = 1.1$) (a) concentration orbit (x, y); (b) rate orbit $(\dot{x}, \dot{y})$; (c) trajectory $(x, \dot{x})$; (d) trajectory $(y, \dot{y})$. Parameters ($a = 0.4$, $b = 1.2$, $Q_0 = 0.03$), (stable: $x_0 \approx 0.450409$, $y_0 \approx 2.410052$; unstable: $x_0 \approx 0.438798$, $y_0 \approx 3.128603$). U and S are for unstable and stable periodic solutions, respectively. Dashed and solid curves are for stable and unstable periodic evolutions, respectively. (*Continues.*)

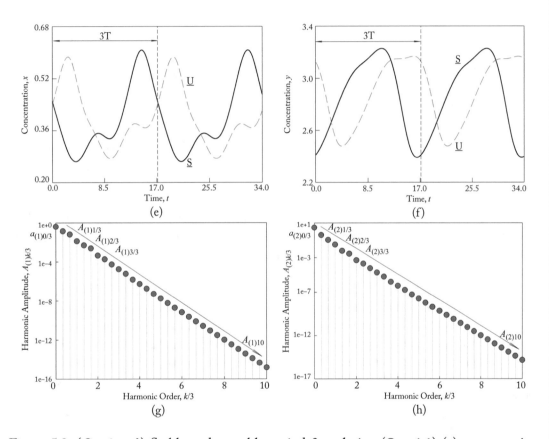

Figure 5.9: (*Continued.*) Stable and unstable period-3 evolution ($\Omega = 1.1$) (e) concentration (t, x); (f) concentration (t, y); (g) harmonic amplitudes of concentration x (stable); (h) harmonic amplitude of concentration y (stable). Parameters ($a = 0.4$, $b = 1.2$, $Q_0 = 0.03$), (stable: $x_0 \approx 0.450409$, $y_0 \approx 2.410052$; unstable: $x_0 \approx 0.438798$, $y_0 \approx 3.128603$). $\underline{U}$ and $\underline{S}$ are for unstable and stable periodic solutions, respectively. Dashed and solid curves are for stable and unstable periodic evolutions, respectively.

Table 5.2: Input data for numerical illustration ($a = 0.4$, $b = 1.2$, $Q_0 = 0.08$)

Ω	Initial Conditions	P-m Evolution
1.1	$(x_0, y_0) \approx (0.450409, 2.410052)$	P-3 (stable)
	$(x_0, y_0) \approx (0.438798, 3.128603)$	P-3 (unstable)
0.97	$(x_0, y_0) \approx (0.380447, 3.221426)$	P-5 (stable)
	$(x_0, y_0) \approx (0.415958, 3.156286)$	P-5 (unstable)
0.92	$(x_0, y_0) \approx (0.422975, 3.109941)$	P-7 (stable)
	$(x_0, y_0) \approx (0.398896, 3.196737)$	P-7 (unstable)
0.9	$(x_0, y_0) \approx (0.360659, 2.994715)$	P-9 (stable)
	$(x_0, y_0) \approx (0.360940, 3.039263)$	P-9 (unstable)

(d), respectively. For the trajectory of $(x, \dot{x})$, the slow-varying portion forms a small swirling cycle near the zero changing rate and the fast-varying portion experiences the large variation of the concentration rate. For the trajectory of $(y, \dot{y})$, the slow-varying portion has an almost constant change rate, and the fast-varying portion has a negative parabolic part and a positive parabolic part.

The time histories of the two concentrations x and y are plotted in Fig. 5.9(e) and (f), respectively. The concentration x for the period-3 motion possesses two peaks connected by a slight wavy portion. Such a wavy portion corresponds to the whirling evolution in the trajectory. The concentration y has an asymmetric parabolic curve in its time history. Left part before the concentration reaches its own maximum varies lower than the right part pass the maximum concentration. Obviously, the two concentrations are positive because of the existence of their corresponding chemicals.

To understand such an independent period-3 evolution, the harmonic amplitudes are very important. The harmonic spectrums of the concentrations x and y for the stable period-3 evolution are presented in Fig. 5.9(g) and (h), respectively. For the concentration x, $a_{(1)0/3} = 0.4$. The main harmonic amplitudes are $A_{(1)1/3} \approx 0.1359$, $A_{(1)2/3} \approx 0.0631$, and $A_{(1)3/3} \approx 0.0128$ for the first three harmonic terms. The quantity of the harmonic amplitude decreases to 10^{-2}. $A_{(1)4/3} \approx 0.0049$, $A_{(1)5/3} \approx 0.0024$, and $A_{(1)6/3} \approx 0.0004$ are for the next three orders. The other harmonic amplitudes are $A_{(1)k/3} \in (10^{-15}, 10^{-3})$ for $k = 7, 8, \ldots, 30$ with $A_{(1)30/3} \approx 1.5 \times 10^{-15}$. For the concentration y, the constant $a_{(2)0/3} \approx 2.8620$. The main harmonic amplitudes are $A_{(2)1/3} \approx 0.3947$, $A_{(2)2/3} \approx 0.1066$, $A_{(2)3/3} \approx 0.0311$, $A_{(2)4/3} \approx 0.0059$, $A_{(2)5/3} \approx 0.0027$, and $A_{(2)6/3} \approx 0.0005$. The other harmonic terms are $A_{(2)k/3} \in (10^{-15}, 10^{-3})$ for $k = 7, 8, \ldots, 30$ with $A_{(1)30/3} \approx 1.5 \times 10^{-15}$, $A_{(1)30/3} \approx 1.5 \times 10^{-15}$. For such an independent period-3 evolution, the harmonic amplitude decreases slowly with harmonic order. The harmonic amplitude of sixth-order term is a level of 10^{-4}. Thus, such a period-3 evolution with the first six harmonic terms will get such an accurate with the error about 10^{-4}. From the harmonic spectrum, the analytical period-3 evolution is about the accuracy of 10^{-15} with 30 harmonic terms. For the unstable period-3 evolution, the harmonic amplitude spectrums are quite similar to the corresponding stable period-3 evolution. Thus, the harmonic amplitudes for the unstable period-3 evolution will not be presented herein.

5.2.2   PERIOD-5 EVOLUTIONS

In Fig. 5.10, the paired stable and unstable period-5 evolutions of the Brusselator are presented for $\Omega = 0.97$. In Fig. 5.10(a), the orbit of the two concentrations is more complicated than that of the period-3 evolution. Two cycles of the period-5 evolutions are observed. The difference between the stable and unstable period-5 evolutions are clearly observed. The rate orbit of the two concentrations are placed in Fig. 5.10(b) for the stable and unstable period-5 evolutions. The trajectories in phase spaces $(x, \dot{x})$ and $(y, \dot{y})$ are presented in Fig. 5.10(c) and (d), respectively. For the trajectory of concentration x, two cycles form a trajectory for the stable or unstable period-5 evolution. There are two whirls near the zero changing rate. The trajectory of the concentration y consists of two big cycles with two small cycles for the stable or unstable period-5 evolution. The slow-varying portions has small rates of concentration, which are close to zero. The difference between the stable and unstable evolutions is observed. With time increase, the unstable period-5 evolution will approach to the stable period-5 evolutions. The time histories of the two concentrations x and y are plotted in Fig. 5.10(e) and (f), respectively. The concentration x for the period-5 motion possesses two peaks. Another wavy portion follows the rapid diminishing form the second peak. The concentration y has two asymmetric parabolic curves, and an almost constant part has a slow-varying rate. Compared to period-3 evolutions, the stable and unpaired stable period-5 evolutions are quite near each other.

To understand the independent stable and unstable period-5 evolution, the harmonic spectrums of the concentrations x and y for the stable period-5 evolution are presented in Fig. 5.10(g) and (h), respectively. For the concentration $x, a_{(1)0/5} = 0.4$. The main harmonic amplitudes are $A_{(1)1/5} \approx 0.0059$, $A_{(1)2/5} \approx 0.0967$, $A_{(1)3/5} \approx 0.0293$, $A_{(1)4/5} \approx 0.0227$, $A_{(1)5/5} \approx 0.0294$, $A_{(1)6/5} \approx 0.0052$, $A_{(1)7/5} \approx 0.0063$, $A_{(1)8/5} \approx 0.0010$, $A_{(1)9/5} \approx 0.0018$, $A_{(1)10/5} \approx 0.0007$. The other harmonic $A_{(1)k/5} \in (10^{-14}, 10^{-3})$ for $k = 11, 12, \ldots, 50$ with $A_{(1)50/5} \approx 1.1 \times 10^{-14}$. For the concentration $y, a_{(2)0/5} \approx 2.9218$. The main harmonic amplitudes are $A_{(2)1/5} \approx 0.0310$, $A_{(2)2/5} \approx 0.2675$, $A_{(2)3/5} \approx 0.0583$, $A_{(2)4/5} \approx 0.0371$, $A_{(2)5/5} \approx 0.0409$, $A_{(2)6/5} \approx 0.0068$, $A_{(2)7/5} \approx 0.0078$, $A_{(2)8/5} \approx 0.0012$, $A_{(2)9/5} \approx 0.0020$, $A_{(2)10/5} \approx 0.0008$. The other harmonic terms are $A_{(2)k/5} \in (10^{-14}, 10^{-3})$ for $k = 11, 12, \ldots, 50$ with $A_{(2)50/5} \approx 1.1 \times 10^{-14}$. The harmonic amplitude varying with harmonic order for the independent stable period-5 evolution is much slower than the independent stable period-3 evolution. The harmonic amplitude of tenth order is of a level 10^{-3}. The period-5 evolution can be approximated at least with ten harmonic terms. From the harmonic spectrum, the analytical period-5

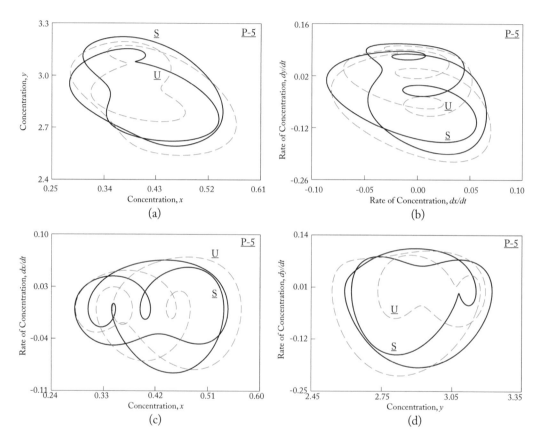

Figure 5.10: Stable and unstable period-5 evolution ($\Omega = 0.97$) (a) concentration orbit (x, y); (b) rate orbit $(\dot{x}, \dot{y})$; (c) trajectory $(x, \dot{x})$; (d) trajectory $(y, \dot{y})$; Parameters ($a = 0.4$, $b = 1.2$, $Q_0 = 0.03$), (stable: $x_0 \approx 0.380447$, $y_0 \approx 3.221426$; unstable: $x_0 \approx 0.415958$, $y_0 \approx 3.156286$). $\underline{U}$ and $\underline{S}$ are for unstable and stable periodic solutions, respectively. Dashed and solid curves are for stable and unstable periodic evolutions, respectively. (*Continues.*)

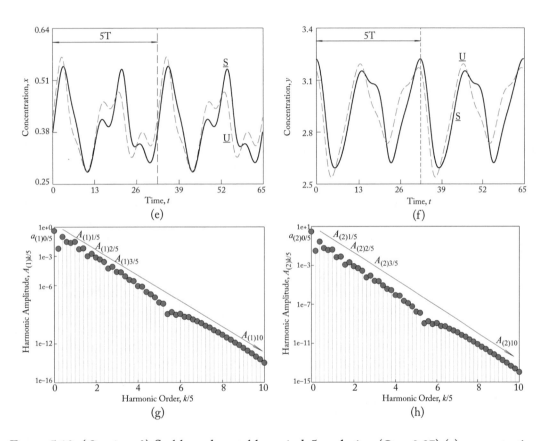

Figure 5.10: (*Continued.*) Stable and unstable period-5 evolution ($\Omega = 0.97$) (e) concentration (t, x); (f) concentration (t, y); (g) harmonic amplitudes of concentration x (stable); (h) harmonic amplitude of concentration y (stable). Parameters ($a = 0.4$, $b = 1.2$, $Q_0 = 0.03$), (stable: $x_0 \approx 0.380447$, $y_0 \approx 3.221426$; unstable: $x_0 \approx 0.415958$, $y_0 \approx 3.156286$). $\underline{U}$ and $\underline{S}$ are for unstable and stable periodic solutions, respectively. Dashed and solid curves are for stable and unstable periodic evolutions, respectively.

evolution is about the accuracy of 10^{-14} with 50 harmonic terms. Again, the harmonic amplitude spectrums for the unstable period-5 evolutions will not be presented herein.

Compared to the independent period-3 evolutions, the stable and unstable period-5 evolutions are much close each other. The unstable and stable period-3 evolutions are different, which can be observed through the solid and dashed curves in the sets of illustrations.

5.2.3 PERIOD-7 EVOLUTIONS

In Fig. 5.11, the stable and unstable period-7 evolutions of the Brusselator are presented for $\Omega = 0.92$. In Fig. 5.11(a), the orbit of the two concentrations is more complicated than that of the period-5 evolution. Three cycles plus a small swirling knot forms a stable or unstable period-7 evolution. The difference between the stable and unstable periodic evolutions are clearly observed. In Fig. 5.11(b), the velocity place of the two concentrations is used for the stable and unstable period-7 evolutions. The velocity trajectory for the stable or unstable period-7 evolution have three large cycles plus three small cycles. The trajectories of the concentrations x and y with the corresponding rates $\dot{x}$ and $\dot{y}$ are presented in Fig. 5.11(c) and (d), respectively. For the trajectory in the plane of $(x, \dot{x})$, three large cycles plus two small swirling cycles near the zero changing rate are observed for the stable and unstable period-7 evolutions. For the trajectory in the plane of $(y, \dot{y})$, two heart-shaped cycles with a small cycle with a swirling knot are observed in Fig. 5.11(e) for the stable and unstable period-7 evolutions. The heart-shaped cycles possess the fast variation portion and slow variation portion. The small cycle is waving near the zero rate of the concentration y. The time histories of the two concentrations x and y are plotted in Fig. 5.11(e) and (f), respectively. The concentration x for the period-7 evolutions possess two positive peaks with one small double-peaks for seven (7) periods. The first two peaks possess similar quantity levels. For the small double-peaks, the concentration varies much slowly compared to the large and sharp peaks. In Fig. 5.11(f), the concentration y has three similar asymmetric waving curves in its time history.

To understand such independent period-7 evolutions, the harmonic spectrums of the concentrations x and y are presented in Fig. 5.11(g) and (h), respectively. For the concentration x, the constant is still $a_{(1)0/7} = 0.4$. The main harmonic amplitudes are $A_{(1)1/7} \approx 0.0059$, $A_{(1)2/7} \approx 0.0053$, $A_{(1)3/7} \approx 0.0877$, $A_{(1)4/7} \approx 0.0344$, $A_{(1)5/7} \approx 0.0032$, $A_{(1)6/7} \approx 0.0193$, $A_{(1)7/7} \approx 0.0321$, $A_{(1)8/7} \approx 0.0021$, $A_{(1)9/7} \approx 0.0036$, $A_{(1)10/7} \approx 0.0065$, $A_{(1)11/7} \approx 0.0022$, $A_{(1)12/7} \approx 0.0005$, $A_{(1)13/7} \approx 0.0018$, $A_{(1)14/7} \approx 0.0007$. The other harmonic $A_{(1)k/7} \in (10^{-13}, 10^{-3})$ for $k = 15, 16, \ldots, 70$ with $A_{(1)70/7} \approx 3.9 \times 10^{-14}$. For the concentration y, the constant is $a_{(2)0/7} \approx 2.9296$. The main harmonic amplitudes are $A_{(2)1/7} \approx 0.0446$, $A_{(2)2/7} \approx 0.0208$, $A_{(2)3/7} \approx 0.2392$, $A_{(2)4/7} \approx 0.0058$, $A_{(2)6/7} \approx 0.0311$, $A_{(2)7/7} \approx 0.0474$, $A_{(2)8/7} \approx 0.0029$, $A_{(2)9/7} \approx 0.0047$, $A_{(2)10/7} \approx 0.0082$, $A_{(2)11/7} \approx 0.0026$, $A_{(2)12/7} \approx 0.0006$, $A_{(2)13/7} \approx 0.0020$, $A_{(2)14/7} \approx 0.0008$. The other harmonic terms are $A_{(2)k/7} \in (10^{-13}, 10^{-3})$ for $k = 15, 16, \ldots, 70$ with $A_{(2)70/7} \approx 3.9 \times 10^{-14}$. For the stable independent period-7 evolution, the harmonic amplitude varying with harmonic order is slow as in the independent period-5 evolu-

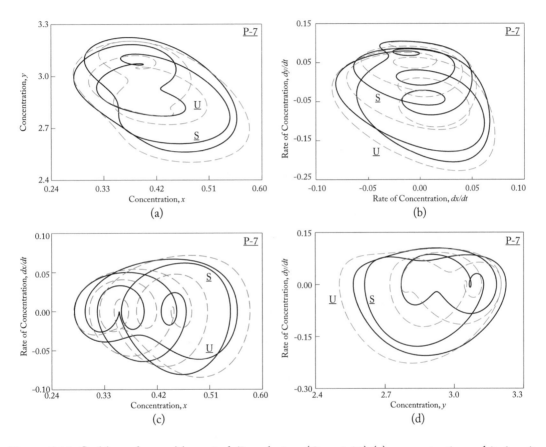

Figure 5.11: Stable and unstable period-7 evolution ($\Omega = 0.92$) (a) concentration orbit (x, y); (b) rate orbit $(\dot{x}, \dot{y})$; (c) trajectory $(x, \dot{x})$; (d) trajectory $(y, \dot{y})$. Parameters ($a = 0.4$, $b = 1.2$, $Q_0 = 0.03$), (stable: $x_0 \approx 0.422975$, $y_0 \approx 3.109941$; unstable: $x_0 \approx 0.398896$, $y_0 \approx 3.196737$). $\underline{U}$ and $\underline{S}$ are for unstable and stable periodic solutions, respectively. Dashed and solid curves are for stable and unstable periodic evolutions, respectively. (*Continues.*)

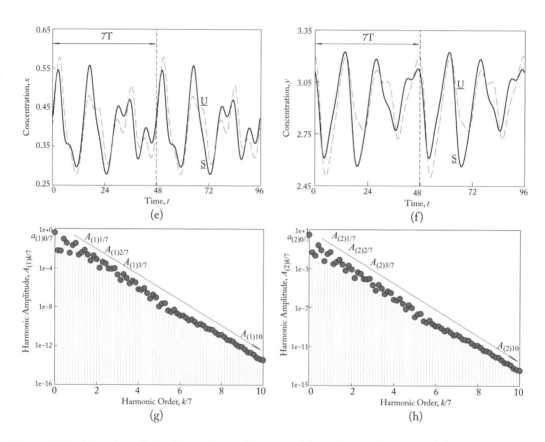

Figure 5.11: (*Continued.*) Stable and unstable period-7 evolution ($\Omega = 0.92$) (e) concentration (t, x); (f) concentration (t, y); (g) harmonic amplitudes of concentration x (stable); (h) harmonic amplitude of concentration y (stable). Parameters ($a = 0.4$, $b = 1.2$, $Q_0 = 0.03$), (stable: $x_0 \approx 0.422975$, $y_0 \approx 3.109941$; unstable: $x_0 \approx 0.398896$, $y_0 \approx 3.196737$). $\underline{U}$ and $\underline{S}$ are for unstable and stable periodic solutions, respectively. Dashed and solid curves are for stable and unstable periodic evolutions, respectively.

tion. The harmonic amplitude of 14th-order term is a quantity level of 10^{-4}. From the harmonic spectrum, the analytical period-7 evolution is about the accuracy of 10^{-14} with 70 terms. The amplitude spectrum for the corresponding unstable period-7 evolution will not be presented herein.

5.2.4   PERIOD-9 EVOLUTIONS

In Fig. 5.12, the stable and unstable period-9 evolutions of the Brusselator are presented for $\Omega = 0.9$. In Fig. 5.12(a), the orbit of the two concentrations (x and y) has four cycles with two twisting small cycles for the stable or unstable period-9 evolution. Compared to period-3, period-5, and period-7 evolutions, the period orbits become more complex. In Fig. 5.12(b), the velocity trajectories of the two concentrations for the stable and unstable period-9 evolutions are presented through the solid and dashed curves. Each of two trajectories have six cycles with two small twisting cycles, and the stable and unstable period-9 evolutions are observed clearly. The trajectories in phase planes of $(x, \dot{x})$ and $(y, \dot{y})$ are presented in Fig. 5.12(c) and (d), respectively. Six cycles plus a twisting cycles formed the a closed curve for the trajectory in the phase plane of $(x, \dot{x})$. The four heart-shaped cycles with a small whirling cycle generates the trajectories in the phase plane of $(y, \dot{y})$, as shown in Fig. 5.12(d). The heart-typed curves possess the slowing variation portions and the fast-varying portions. The time histories of the two concentrations (x and y)are plotted in Fig. 5.12(e) and (f), respectively. The concentration x for the period-9 motion endures four positive single peaks in nine (9) periods. The entire time history response of the concentration x has two single-peaks and two double-peaks. In Fig. 5.12(f), the time response of concentration y has four asymmetric portions in its time history. The entire response looks like a beating oscillation with four fast-waving oscillations. The stable and unstable period-9 evolutions are quite close each other.

To understand such an independent period-9 evolution, the harmonic spectrums of the concentrations x and y are presented in Fig. 5.12(g) and (h), respectively. For the concentration x, the constant is still $a_{(1)0/9} = 0.4$. The main harmonic amplitudes are $A_{(1)1/9} \approx 0.0047$, $A_{(1)2/9} \approx 0.0017$, $A_{(1)3/9} \approx 0.0090$, $A_{(1)4/9} \approx 0.0803$, $A_{(1)5/9} \approx 0.0370$, $A_{(1)6/9} \approx 0.0043$, $A_{(1)7/9} \approx 0.0038$, $A_{(1)8/9} \approx 0.0165$, $A_{(1)9/9} \approx 0.0333$, $A_{(1)10/9} \approx 0.0024$, $A_{(1)11/9} \approx 0.0011$, $A_{(1)12/9} \approx 0.0029$, $A_{(1)13/9} \approx 0.0062$, $A_{(1)14/9} \approx 0.0027$, $A_{(1)15/9} \approx 0.0001$, $A_{(1)16/9} \approx 0.0005$, $A_{(1)17/9} \approx 0.0016$, $A_{(1)18/9} \approx 0.0008$. The other harmonic amplitudes are $A_{(1)k/9} \in (10^{-13}, 10^{-3})$ for $k = 19, 20, \ldots, 90$ with $A_{(1)90/9} \approx 4.2 \times 10^{-14}$. For the concentration y, the constant is $a_{(2)0/9} \approx 2.9358$. The main harmonic amplitudes are $A_{(2)1/9} \approx 0.0473$, $A_{(2)2/9} \approx 0.0086$, $A_{(2)3/9} \approx 0.0314$, $A_{(2)4/9} \approx 0.2163$, $A_{(2)5/9} \approx 0.0827$, $A_{(2)6/9} \approx 0.0085$, $A_{(2)7/9} \approx 0.0066$, $A_{(2)8/9} \approx 0.0265$, $A_{(2)9/9} \approx 0.0498$, $A_{(2)10/9} \approx 0.0034$, $A_{(2)11/9} \approx 0.0014$, $A_{(2)12/9} \approx 0.0038$, $A_{(2)13/9} \approx 0.0079$, $A_{(2)14/9} \approx 0.0034$, $A_{(2)15/9} \approx 0.0001$, $A_{(2)16/9} \approx 0.0006$, $A_{(2)17/9} \approx 0.0018$, $A_{(2)18/9} \approx 0.0009$. The other harmonic terms are $A_{(2)k/9} \in (10^{-13}, 10^{-3})$ for $k = 19, 20, \ldots, 90$ with $A_{(2)90/9} \approx 4.2 \times 10^{-14}$. For this independent period-9 evolution, the harmonic amplitude varying with harmonic order is very

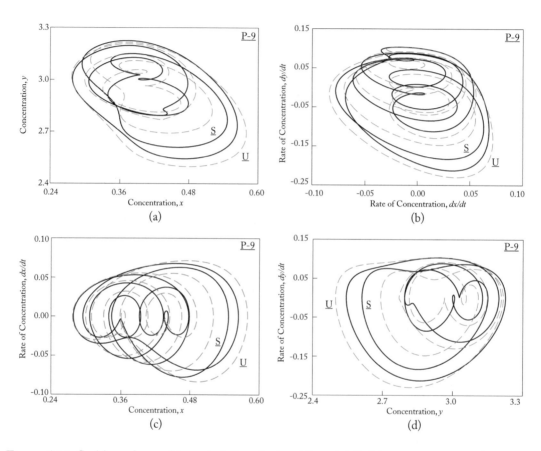

Figure 5.12: Stable and unstable period-9 evolution ($\Omega = 0.90$): (a) concentration orbit (x, y); (b) rate orbit $(\dot{x}, \dot{y})$; (c) trajectory $(x, \dot{x})$; (d) trajectory $(y, \dot{y})$. Parameters ($a = 0.4$, $b = 1.2$, $Q_0 = 0.03$), (stable: $x_0 \approx 0.360659$, $y_0 \approx 2.994715$; unstable: $x_0 \approx 0.360940$, $y_0 \approx 3.039263$). $\underline{U}$ and $\underline{S}$ are for unstable and stable periodic solutions, respectively. Dashed and solid curves are for stable and unstable periodic evolutions, respectively. (*Continues.*)

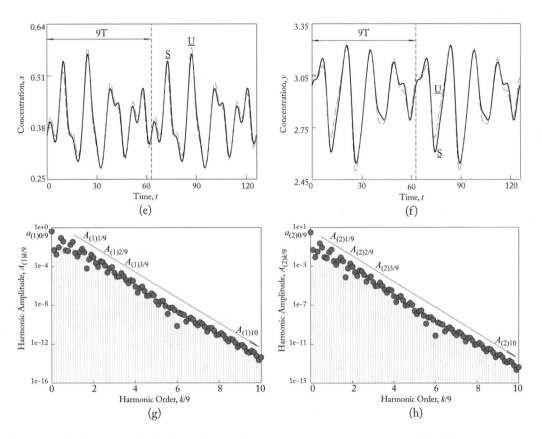

Figure 5.12: (*Continued.*) Stable and unstable period-9 evolution ($\Omega = 0.90$): (e) concentration (t, x); (f) concentration (t, y); (g) harmonic amplitudes of concentration x (stable); (h) harmonic amplitude of concentration y (stable). Parameters ($a = 0.4$, $b = 1.2$, $Q_0 = 0.03$), (stable: $x_0 \approx 0.360659$, $y_0 \approx 2.994715$; unstable: $x_0 \approx 0.360940$, $y_0 \approx 3.039263$). $\underline{U}$ and $\underline{S}$ are for unstable and stable periodic solutions, respectively. Dashed and solid curves are for stable and unstable periodic evolutions, respectively.

slow. The harmonic amplitude of the 18th-order term is a quantity level of 10^{-4}. Such a stable period-9 evolution needs 18 harmonic terms to be approximated. From the harmonic spectrum, the analytical period-9 evolution is about the accuracy of 10^{-14} with 90 harmonic terms. For the corresponding unstable period-9 evolution, the similar harmonic spectrum can be presented from the analytical solutions.

CHAPTER 6

Production and Compensation

Without loss of generality, assume $k_i = 1$ $(i = 1, 2, \ldots, 4)$ with discussion. The catalysts X and Y are produced and depleted simultaneously in the Brusselator. Due to the effect of combination, the concentration of catalysts X and Y will fluctuate in the entire reaction duration. The average concentration of X and Y can be obtained by integration, i.e.,

$$\varpi_X = \lim_{t^* \to \infty} \frac{1}{t^*} \int_0^{t^*} x \, dt = a_{(1)0/m},$$

$$\varpi_Y = \lim_{t^* \to \infty} \frac{1}{t^*} \int_0^{t^*} y \, dT = a_{(2)0/m},$$

(6.1)

where ϖ_X and ϖ_Y are average concentration of X and Y, respectively. Because the compensation of reactants A and B should equal to their consumption as in Eq. (3.2) to keep them constant, their average compensation are

$$C_A = \lim_{t^* \to \infty} \frac{1}{t^*} \int_0^{t^*} -k_1[A] dT = \lim_{t^* \to \infty} \frac{1}{t^*} \int_0^{t^*} -a \, dt = -a,$$

$$C_B = \lim_{t^* \to \infty} \frac{1}{t^*} \int_0^{t^*} -k_2[B][X] dT = \lim_{t^* \to \infty} \frac{1}{t^*} \int_0^{t^*} -bx \, dt = -ba_{(1)0/m},$$

(6.2)

where C_A and C_B are consumption rate of reactants A and B, respectively. The average productivity is given by integrating the governing equation in Eq. (3.2) by time t, i.e.,

$$P_D = \lim_{t^* \to \infty} \frac{1}{t^*} \int_0^{t^*} k_2[B][X] dT = \lim_{t^* \to \infty} \frac{1}{t^*} \int_0^{t^*} bx \, dt = ba_{(1)0/m},$$

$$P_E = \lim_{t^* \to \infty} \frac{1}{t^*} \int_0^{t^*} k_4[X] dT = \lim_{t^* \to \infty} \frac{1}{t^*} \int_0^{t^*} x \, dt = a_{(1)0/m}.$$

(6.3)

Substitution of Eq. (6.1) into Eqs. (6.2) and (6.3) yields

$$P_E = -C_A = a,$$
$$P_D = -C_B = ab,$$
$$\frac{C_A}{C_B} = \frac{P_E}{P_D} = \frac{1}{b}.$$

(6.4)

In terms of compensation and production, the evolution relation of the Brusselator can be understood. For simply speaking, the reactant A produces product E directly, and the reactant B produces product D. In the Brusselator, the proportion of the two products D and E is totally determined by the concentration of reactant B if the rate constants are given. Controlling the concentration of reactant B is feasible for the modern chemical industry. It implies that engineers can design any product ratio through changing the concentration of reactant B.

References

[1] Prigogine, I. and Lefever, R. (1968). Symmetry breaking instabilities in dissipative systems. II, *The Journal of Chemical Physics*, 48(4):1695–1700. DOI: 10.1063/1.1668896. 1, 15

[2] Lefever, R. and Nicolis, G. (1971). Chemical instabilities and sustained oscillations, *Journal of Theoretical Biology*, 30(2):267–284. DOI: 10.1016/0022-5193(71)90054-3. 1

[3] Tyson, J. J. (1973). Some further studies of nonlinear oscillations in chemical systems, *The Journal of Chemical Physics*, 58(9):3919–3930. DOI: 10.1063/1.1679748. 1

[4] Tomita, K., Kai, T., and Hikami, F. (1977). Entrainment of a limit cycle by a periodic external excitation, *Progress of Theoretical Physics*, 57(4):1159–1177. DOI: 10.1143/ptp.57.1159. 1, 17

[5] Hao, B. L. and Zhang, S. Y. (1982). Hierarchy of chaotic bands, *Journal of Statistical Physics*, 28(4):769–792. DOI: 10.1007/bf01011880. 1

[6] Roy, T., Choudhury, R., and Tanriver, U. (2017). Analytical prediction of homoclinic bifurcations following a supercritical Hopf bifurcation, *Discontinuity, Nonlinearity, and Complexity*, 6(2):209–222. DOI: 10.5890/dnc.2016.09.002. 1

[7] Maaita, J. O. (2016). A theorem on the bifurcations of the slow invariant manifold of a system of two linear oscillators coupled to a k-order nonlinear oscillator, *Journal of Applied Nonlinear Dynamics*, 5(2):193–197. DOI: 10.5890/jand.2016.06.006. 1

[8] Yamgoué, S. B., Nana, B., and Pelap, F. B. (2017). Approximate analytical solutions of a nonlinear oscillator equation modeling a constrained mechanical system, *Journal of Applied Nonlinear Dynamics*, 6(1):17–26. DOI: 10.5890/jand.2017.03.002. 1

[9] Shayak, B. and Vyas, P. (2017). Krylov Bogoliubov type analysis of variants of the Mathieu equation, *Journal of Applied Nonlinear Dynamics*, 6(1):57–77. DOI: 10.5890/jand.2017.03.005. 1

[10] Rajamani, S. and Rajasekar, S. (2017). Variation of response amplitude in parametrically driven single Duffing oscillator and unidirectionally coupled Duffing oscillators, *Journal of Applied Nonlinear Dynamics*, 6(1):121–129. DOI: 10.5890/jand.2017.03.009. 1

[11] Cochelin, B. and Vergez, C. (2009). A high order purely frequency-based harmonic balance formulation for continuation of periodic solutions, *Journal of Sound and Vibration*, 324(1–2):243–262. DOI: 10.1016/j.jsv.2009.01.054. 1

[12] Luo, A. C. J. (2012). *Continuous Dynamical Systems*, HEP/L&H, Beijing/Glen Garbon. 1, 3, 5, 11, 14, 17, 21

[13] Luo, A. C. J. and Huang, J. (2012). Approximate solutions of periodic motions in nonlinear systems via a generalized harmonic balance, *Journal of Vibration and Control*, 18(11):1661–1674. DOI: 10.1177/1077546311421053. 1

[14] Luo, A. C. J. and Huang, J. Z. (2012). Analytical dynamics of period-*m* flows and chaos in nonlinear Systems, *International Journal of Bifurcation and Chaos*, 22(4):29. (Article no. 1250093). DOI: 10.1142/s0218127412500939. 1

[15] Luo, A. C. J. and Huang, J. Z. (2012). Analytical routes of period-1 motions to chaos in a periodically forced Duffing oscillator with a twin-well potential, *Journal of Applied Nonlinear Dynamics*, 1:73–108. DOI: 10.5890/jand.2012.02.002. 2

[16] Luo, A. C. J. and Huang, J. Z. (2012). Unstable and stable period-*m* motions in a twin-well potential Duffing oscillator, *Discontinuity, Nonlinearity and Complexity*, 1:113–145. DOI: 10.5890/dnc.2012.03.001. 2

[17] Luo, A. C. J. and Huang, J. Z. (2013). Analytical solutions for asymmetric periodic motions to chaos in a hardening Duffing oscillator, *Nonlinear Dynamics*, 72:417–438. DOI: 10.1007/s11071-012-0725-3. 2

[18] Luo A. C. J. and Huang, J. Z. (2013). Analytical period-3 motions to chaos in a hardening Duffing oscillator, *Nonlinear Dynamics*, 73:1905–1932. DOI: 10.1007/s11071-013-0913-9. 2, 73

[19] Luo, A. C. J. and Huang, J. Z. (2013). An analytical prediction of period-1 motions to chaos in a softening Duffing oscillator, *International Journal of Bifurcation and Chaos*, 23(5):31. (Article no. 1350086). DOI: 10.1142/S0218127413500867. 2, 73

[20] Luo, A. C. J. and Huang, J. Z.(2014). Period-3 motions to chaos in a softening Duffing oscillator, *International Journal of Bifurcation and Chaos*, 24:26. (Article no. 1430010). DOI: 10.1142/s0218127414300109. 2

[21] Luo, A. C. J. and Yu, B. (2013). Complex period-1 motions in a periodically forced, quadratic nonlinear oscillator, *Journal of Vibration of Control*, 21(5):907–918. DOI: 10.1177/1077546313490525. 2, 21

[22] Luo, A. C. J. and Lakeh, A. B. (2014). An approximate solution for period-1 motions in a periodically forced van der Pol oscillator, ASME, *Journal of Computational and Nonlinear Dynamics*, 9(3):7. (Article no. 031001). DOI: 10.1115/1.4026425. 2, 21

[23] Luo, A. C. J. and Lakeh, A. B. (2013). Analytical solution for period-*m* motions in a periodically forced, van der Pol oscillator, *International Journal of Dynamics and Control*, 1(2):99–115. DOI: 10.1115/imece2012-86589. 2

[24] Luo, A. C. J. and Lakeh, A. B. (2014). Period-*m* motions and bifurcation in a periodically forced van der Pol-Duffing oscillator, *International Journal of Dynamics and Control*, 2(4):474–493. DOI: 10.1007/s40435-014-0058-9. 2

[25] Luo, A. C. J. and Yu, B. (2015). Bifurcation trees of period-1 motions to chaos in a two-degree-of-freedom, nonlinear oscillator, *International Journal of Bifurcation and Chaos*, 25(1):3. (Article no. 1550179, 40 pages). DOI: 10.1142/s0218127415501795. 2

[26] Yu, B. and Luo, A. C. J. (2017). Analytical period-1 motions to chaos in a two-degree-of-freedom oscillator with a hardening nonlinear spring, *International Journal of Dynamics and Control*, 5(3):436–453. DOI: 10.1007/s40435-015-0216-8. 2

[27] Luo, A. C. J. (2013). Analytical solutions for periodic motions to chaos in nonlinear systems with/without time-delay, *International Journal of Dynamics and Control*, 1:330–359. DOI: 10.1007/s40435-013-0024-y. 2

[28] Luo, A. C. J. and Jin, H. X. (2014). Bifurcation trees of period-*m* motion to chaos in a time-delayed, quadratic nonlinear oscillator under a periodic excitation, *Discontinuity, Nonlinearity, and Complexity*, 3:87–107. DOI: 10.5890/dnc.2014.03.007. 2

[29] Luo, A. C. J. and Jin, H. X. (2014). Period-*m* motions to chaos in a periodically forced, Duffing oscillator with a time-delayed displacement, *International Journal of Bifurcation and Chaos*, 24(10):20. (Article no. 1450126). DOI: 10.1142/s0218127414501260. 2

[30] Luo, A. C. J. and Jin, H. X. (2015). Complex period-1 motions of a periodically forced Duffing oscillator with a time-delay feedback, *International Journal of Dynamics and Control*, 3:325–340. DOI: 10.1007/s40435-014-0091-8. 2

[31] Luo, A. C. J. and Jin, H. X. (2015). Period-3 motions to chaos in a periodically forced Duffing oscillator with a linear time-delay, *International Journal of Dynamics and Control*, 3:371–388. DOI: 10.1007/s40435-014-0116-3. 2

[32] Wang, Y. F. and Liu, Z. W. (2015). A matrix-based computational scheme of generalized harmonic balance method for periodic solutions of nonlinear vibratory systems, *Journal of Applied Nonlinear Dynamics*, 4(4):379–389. DOI: 10.5890/jand.2015.11.005. 2

[33] Luo, H. and Wang, Y. F. (2016). Nonlinear dynamics analysis of a continuum rotor through generalized harmonic balance method, *Journal of Applied Nonlinear Dynamics*, 5(1):1–31. DOI: 10.5890/jand.2016.03.001. 2

[34] Akhmet, M. and Fen, M. O. (2018). Almost periodicity in chaos, *Discontinuity, Nonlinearity, and Complexity*, 7(1):15–19. DOI: 10.5890/dnc.2018.03.002. 2

[35] Luo, A. C. J. (2014). *Toward Analytical Chaos in Nonlinear Systems*, Wiley, New York. DOI: 10.1002/9781118887158. 2, 3, 5, 9, 11, 14, 17

[36] Luo, A. C. J. (2014). *Analytical Routes to Chaos in Nonlinear Engineering*, Wiley, New York. DOI: 10.1002/9781118883938. 2, 3, 5, 9, 11, 14, 17

[37] Luo, A. C. J. and Guo, S. (2018). Analytical solutions of period-1 to period-2 motions in a periodically diffused Brusselator, ASME, *Journal of Computational and Nonlinear Dynamics*, 13(9):090912. DOI: 10.1115/1.4038204. 2

[38] Luo, A. C. J. and Guo, S. (2018). Period-1 evolutions to chaos in a periodically forced Brusselator, *International Journal of Bifurcation and Chaos*, 28(14):1830046. DOI: 10.1142/s021812741830046x. 2, 25

[39] Luo, A. C. J. and Guo, S. (2018). On independent period-m evolutions in a periodically forced Brusselator, *Journal of Vibration Testing and Systems Dynamics*, 2(4):375–402. DOI: 10.5890/jvtsd.2018.12.004. 2, 53

Authors' Biographies

ALBERT C.J. LUO

Professor Albert C.J. Luo works at Southern Illinois University Edwardsville. For over 30 years, Dr. Luo's contributions on nonlinear dynamical systems and mechanics lie in: (i) the local singularity theory for discontinuous dynamical systems; (ii) dynamical systems synchronization; (iii) analytical solutions of periodic and chaotic motions in nonlinear dynamical systems; (iv) the theory for stochastic and resonant layer in nonlinear Hamiltonian systems; and (v) the full nonlinear theory for a deformable body. Such contributions have been scattered into 20 monographs and over 300 peer-reviewed journal and conference papers. Dr. Luo has served as an editor for the journal *Communications in Nonlinear Science and Numerical Simulation*, and book series on Nonlinear Physical Science (HEP) and Nonlinear Systems and Complexity (Springer). Dr. Luo was an editorial member for *IMeChE Part K Journal of Multibody Dynamics* and *Journal of Vibration and Control*, and has also organized over 30 international symposiums and conferences on dynamics and control.

SIYU GUO

Siyu Guo is a Ph.D. student in Mechanical Engineering, at Southern Illinois University Edwardsville. He received his B.S. from Nanjing University of Science and Technology in 2012 and his M.S. from Southern Illinois University Edwardsville in 2018. In 2017–2018, he was an FEA engineer in the Innovation & Technology Development Division of the Caterpillar Inc. in Peoria, Illinois. His research interests are on dynamical systems and discontinuous dynamical systems. Siyu has published five peer-review journal papers and five conference papers on nonlinear dynamics.

Printed in the United States
by Baker & Taylor Publisher Services